STUDENT SOLUTIONS MANUAL

Nancy J. Gardner
California State University, Long Beach

Introduction to
CHEMICAL PRINCIPLES
Eleventh Edition

H. Stephen Stoker

PEARSON

Boston   Columbus   Indianapolis   New York   San Francisco   Upper Saddle River
Amsterdam   Cape Town   Dubai   London   Madrid   Milan   Munich   Paris   Montréal   Toronto
Delhi   Mexico City   São Paulo   Sydney   Hong Kong   Seoul   Singapore   Taipei   Tokyo

Editor in Chief: Adam Jaworski
Senior Marketing Manager: Jonathan Cottrell
Assistant Editor: Coleen Morrison
Editorial Assistant: Fran Falk
Managing Editor, Chemistry and Geosciences: Gina M. Cheselka
Project Manager: Meghan DeMaio
Editorial Production and Composition Services: Sudip Sinha/PreMediaGlobal

Cover Photo Credit: Shilova Ekaterina/Shutterstock

PEARSON   www.pearsonhighered.com   ISBN-10: 0-321-81512-2; ISBN-13: 978-0-321-81512-5

Contents

CHAPTER ONE
The Science of Chemistry

SCIENTIFIC DISCIPLINES (SEC. 1.1)

1.1 a) true b) false c) false d) true

1.3 a) false b) false c) true d) false

SCIENTIFIC RESEARCH AND TECHNOLOGY (SEC. 1.2)

1.5 a) basic research b) applied research c) basic research d) applied research

THE SCIENTIFIC METHOD (SEC. 1.4)

1.7 c, b, e, a, and d

1.9 a) scientific hypothesis b) nonscientific hypothesis
c) scientific hypothesis d) nonscientific hypothesis

1.11 a) scientific law b) scientific hypothesis
c) scientific law d) scientific hypothesis

1.13 a) correct b) incorrect
c) correct d) incorrect

1.15 While a scientific theory may not be an absolute answer, it is the best answer available. It may be supplanted only if repeated experimental evidence conclusively disproves it and a new theory is developed.

1.17 a) 4 is eliminated b) 1 is eliminated
c) 1 and 4 are eliminated d) 1 and 2 are eliminated

1.19 a) qualitative b) qualitative
c) quantitative d) quantitative

1.21 The product of the pressure times the volume is a constant; or the pressure of the gas is inversely proportional to its volume; $P_1 V_1 = P_2 V_2$.

1.23 Scientific laws are discovered by research. Researchers have no control over what the laws turn out to be. Societal laws are arbitrary conventions agreed upon (in a democracy) by the majority of those to whom the law applies that can be and are changed by society when necessary.

1.25 Publishing scientific data provides access to that data, enabling scientists to develop new scientific theories based on a wider range of knowledge relating to a particular field.

1.27   Conditions under which an experiment is conducted often affect the results of the experiment. If the conditions are uncontrolled, the data resulting from that experiment are not validated.

1.29   A qualitative observation involves general nonnumerical information about a system under study. A quantitative observation involves numerical measurements.

Answers to Multiple-Choice Practice Test

| **MC 1.1** | e | **MC 1.2** | a | **MC 1.3** | b | **MC 1.4** | d | **MC 1.5** | a |
| **MC 1.6** | d | **MC 1.7** | b | **MC 1.8** | b | **MC1.9** | b | **MC1.10** | c |

CHAPTER TWO
Numbers from Measurements

Practice Problems by Topic

EXACT AND INEXACT NUMBERS (SEC. 2.2)

2.1 a) exact (counting integer)   b) exact (definition)
c) inexact (measurement)   d) exact (counting integer)

2.3 60 seconds in a minute represents an exact number; 60 feet long represents a measured value.

ACCURACY AND PRECISION (SEC. 2.3)

2.5 Student A: low precision, low accuracy   Student B: high precision, high accuracy
Student C: high precision, low accuracy

2.7 a) IV   b) I   c) II   d) III

UNCERTAINTY IN MEASUREMENTS (SEC. 2.4)

2.9 a) Ruler 1 $= \pm 0.1$ cm   b) Ruler 2 $= \pm 0.01$ cm

2.11 a) Ruler 3 $= 27$ cm   b) Ruler 4 $= 27.0$ cm

2.13 a) Ruler 4   b) Ruler 1   c) Ruler 2   d) Ruler 3

2.15 a) ± 0.01 b) ± 0.01 c) ± 0.001 d) ± 0.0001

2.17 The uncertainty in 3.3 seconds is ± 0.1 second and the uncertainty in 3.30 seconds is ± 0.01 second.

2.19 a) 40,000–60,000 people   b) 49,000–51,000 people
c) 49,900–50,100 people   d) 49,990–50,010 people

SIGNIFICANT FIGURES (SEC. 2.5)

2.21 a) 5   b) 4   c) 6   d) 1

2.23 a) 3   b) 6   c) 3   d) 6

2.25 a) 6   b) 6   c) 3   d) 2

2.27 a) confined zeros $= 2$, leading zeros $= 0$, significant trailing zeros $= 1$, trailing zeros not significant $= 0$
b) confined zeros $= 2$, leading zeros $= 3$, significant trailing zeros $= 2$, trailing zeros not significant $= 0$
c) confined zeros $= 1$, leading zeros $= 0$, significant trailing zeros $= 0$, trailing zeros not significant $= 2$
d) confined zeros $= 0$, leading zeros $= 0$, significant trailing zeros $= 0$, trailing zeros not significant $= 6$

2.29 a) 3010.2**0**, **0** is the estimated digit   b) 0.0030030**0**, **0** is the estimated digit
c) 4**0**,400, **4** is the estimated digit   d) **3**3,000,000, **3** is the estimated digit

2.31   a) ±0.01   b) ±0.00000001   c) ±100   d) ±1,000,000

2.33

		Number of Significant Figures	Estimated Digit in Measurement	Magnitude of Uncertainty
	6.75	3	5	±0.01
a)	3.130	4	0	±0.001
b)	0.002	1	2	±0.001
c)	3050	3	5	±10
d)	2100	2	1	±100

2.35   a) same number of significant figures   b) different number of significant figures
            c) same number of significant figures   d) same number of significant figures

2.37   a) same uncertainty, ±0.01   b) different uncertainties, ±1, ±10
            c) same uncertainty, ±0.00001   d) different uncertainties, ±0.000001, ±10,000

2.39   a) 23,000   b) 23,$\overline{0}$ 00   c) 23,000.0   d) 23,000.000

2.41   a) 3   b) 4   c) 2   d) 5

ROUNDING OFF (SEC. 2.6)

2.43   a) 0.0123   b) 0.126   c) 6.24   d) 24.9

2.45   a) 42.6   b) 42.6   c) 42.8   d) 42.8

2.47   a) 42,300   b) 42,400   c) 42,500   d) 42,600

2.49

	Measured Value	4 Significant Figures Value for Measurement	2 Significant Figures Value for Measurement
	2.3552	2.355	2.4
a)	1.7747	1.775	1.8
b)	23,511	23,510	24,000
c)	0.70871	0.7087	0.71
d)	4.3500	4.350	4.4

2.51   a) 0.351   b) 653.9   c) 22.556   d) 0.2777

SIGNIFICANT FIGURES IN MULTIPLICATION AND DIVISION (SEC. 2.6)

2.53   a) 2   b) 1   c) 3   d) 3

2.55   a) 3   b) 2   c) 4 or more   d) 4 or more

2.57   a) 0.029889922 (calc), 0.0299 (corr)   b) 136,900 (calc), 140,000 (corr)
            c) 1277.2522 (calc), 1280 (corr)   d) 0.98816568 (calc), 0.988 (corr)

2.59   a) 3.9265927 (calc), 3.9 (corr)   b) 4.8410309 (calc), 4.84 (corr)
           c) 63.13492 (calc), 63 (corr)   d) 1.1851852 (calc), 1 (corr)

SIGNIFICANT FIGURES IN ADDITION AND SUBTRACTION (SEC. 2.6)

2.61   a) tenths, ± 0.1 b) tenths, ± 0.1
           c) ones, ± 1 d) hundreds, ± 100

2.63   a) 162 (calc and corr)   b) 9.321 (calc), 9.3 (corr)
           c) 1260.72 (calc), 1261 (corr) d) 19.95 (calc), 20.0 (corr)

2.65   a) 957 (calc), 957.0 (corr)   b) 342.63 (calc), 343 (corr)
           c) 1250 (calc), 1200 (corr)   d) 131.9927 (calc), 132 (corr)

CALCULATIONS INVOLVING BOTH SIGNIFICANT FIGURE RULES (SEC. 2.6)

2.67   a) 2.732 (calc), 2.7 (corr)   b) 2.8521 (calc), 2.9 (corr)
           c) 0.219 (calc), 0.2 (corr)   d) 46.393464 (calc), 46.4 (corr)

SIGNIFICANT FIGURES AND EXACT NUMBERS (SEC. 2.6)

2.69   a) 267.3 (calc and corr)   b) 257,140 (calc), 260,000 (corr)
           c) 201.3 (calc and corr)   d) 3.8038461 (calc), 3.8 (corr)

2.71   a) 11×2.21 in. $= 24.31$ in. (calc), 24.3 in. (corr)
           b) $(2.21 + 2.21 + 2.21 + 2.21 + 2.21 + 2.21 + 2.21 + 2.21 + 2.21 + 2.21 + 2.21)$ in.
              $= 24.31$ in. (calc and corr)

SCIENTIFIC NOTATION (SEC. 2.7)

2.73   a) negative   b) positive   c) zero   d) positive

2.75   a) Two decimal places to the right   b) Four decimal places to the right
           c) Three decimal places to the left   d) Four decimal places to the left

2.77   a) 4 significant figures   b) 5 significant figures
           c) 4 significant figures   d) 6 significant figures

2.79   a) 2 digits   b) 5 digits   c) 2 digits   d) 7 digits

2.81   a) 3.25×10^{-2} b) 3.4500×10^{2} c) 3.2×10^{4} d) 9.70003×10^{7}

2.83   a) 6.67×10^{3} b) 3.75×10^{1} c) 3.46×10^{-1} d) 1.30×10^{0}

2.85   a) 7×10^{4} b) 6.70×10^{4} c) 6.7000×10^{4} d) 6.700000×10^{4}

2.87   a) 0.00230   b) 4350   c) 0.066500   d) 111,000,000

2.89

	Decimal Form	Scientific Notation Form	Significant Figures Decimal Form	Significant Figures Scientific Notation Form
	27.23	2.723×10^1	four	four
a)	0.633	6.33×10^{-1}	three	three
b)	33,000	3.3×10^4	two	two
c)	0.1140	1.140×10^{-1}	four	four
d)	63,300	$6.33 \times 10^{4*}$	three	three

*(Note typo in text of 6.33×10^{44} should have been 6.33×10^4).

2.91   a) 3.42×10^6 b) 2.36×10^{-3} c) 3.2×10^2 d) 1.2×10^{-4}

2.93   a) smaller   b) smaller   c) larger   d) larger

UNCERTAINTY AND SCIENTIFIC NOTATION (SEC. 2.7)

2.95   a) $10^{-3} \times 10^4 = \pm 10$ b) $10^{-3} \times 10^6 = \pm 10^3$
            c) $10^{-2} \times 10^5 = \pm 10^3$ d) $10^{-1} \times 10^{-2} = \pm 10^{-3}$

2.97   a) 3.65×10^5 b) 3.6500×10^5 c) 3.650000×10^5 d) 3.65000000×10^5

2.99   a) less than   b) same   c) same   d) greater than

MULTIPLICATION AND DIVISION IN SCIENTIFIC NOTATION (SEC. 2.8)

2.101   a) $\pm 10^0$ b) 2.7×10^3 c) 2.735000×10^3 d) 3×10^3

2.103

	Decimal Form	Scientific Notation Form	Uncertainty Decimal Form	Uncertainty Scientific Notation Form
	27.23	$2.723 \times 10^{1*}$	$\pm 10^{-2}$	10^{-2}
a)	0.6770	6.770×10^{-1}	$\pm 10^{-4}$	10^{-4}
b)	33,003	3.3003×10^4	$\pm 10^0$	10^0
c)	635.0	6.350×10^2	$\pm 10^{-1}$	10^{-1}
d)	0.06350	6.350×10^{-2}	$\pm 10^{-5}$	10^{-5}

*(Note typo in text of 2.723×10^{11} should have been 2.723×10^1).

2.105   a) 10^8 b) 10^{-8} c) 10^2 d) 10^{-2}

2.107   a) 2.991905×10^8 (calc), 2.992×10^8 (corr)   b) 9.129×10^2 (calc), 9.1×10^2 (corr)
             c) 2.7×10^{-9} (calc and corr)   d) 2.7×10^{11} (calc and corr)

2.109   a) 10^2 b) 10^8 c) 10^{-8} d) 10^{-2}

2.111   a) 2.8649608×10^1 (calc), 2.86×10^1 (corr)   b) 8.9991001×10^{17} (calc), 8.999×10^{17} (corr)
             c) 3.490449×10^{-2} (calc), 3.49×10^{-2} (corr)   d) $1.1112222 \times 10^{-18}$ (calc), 1.111×10^{-18} (corr)

2.113 a) 10^1 b) 10^{-1} c) 10^{20} d) 10^2

2.115 a) 1.5×10^0 (calc and corr) b) 6.6666666×10^{-1} (calc), 6.7×10^{-1} (corr)
c) $8.5073917 \times 10^{-19}$ (calc), 8.51×10^{-19} (corr) d) 7.7775×10^{15} (calc), 8×10^{15} (corr)

ADDITION AND SUBTRACTION IN SCIENTIFIC NOTATION (SEC. 2.8)

2.117 a) 4.415×10^3 (calc), 4.42×10^3 (corr) b) 9.3×10^{-2} (calc), 9.30×10^{-2} (corr)
c) 9.683×10^5 (calc and corr) d) 1.9189×10^4 (calc), 1.919×10^4 (corr)

2.119 a) 7.713×10^7 (calc and corr) b) 8.253×10^7 (calc and corr)
c) 8.307×10^7 (calc and corr) d) 8.31294×10^7 (calc), 8.313×10^7 (corr)

2.121 a) 4.415×10^3 (calc), 4.42×10^3 (corr) b) 9.3×10^{-2} (calc), 9.30×10^{-2} (corr)
c) 9.683×10^5 (calc and corr) d) 1.9189×10^4 (calc), 1.919×10^4 (corr)

2.123 a) 7.713×10^7 (calc and corr) b) 8.253×10^7 (calc and corr)
c) 8.307×10^7 (calc and corr) d) 8.31294×10^7 (calc), 8.313×10^7 (corr)

Multi-Concept Problems

2.125 a) yes b) no c) yes d) no

2.127 a) 3 b) 4 c) 4 d) 4

2.129 a) no b) yes c) yes d) yes

2.131 a) yes b) no c) no d) no

2.133 a) 2 b) 4 or more c) 3 d) 3

2.135 An exact number is a whole number; it cannot possess decimal digits.

2.137 a) $4.2 + 5.30 = 9.5$ (calc and corr)
$28 + 11 = 39$ (exact)
$39 \times 9.5 = 370.5$ (calc), 3.7×10^2 (corr)
b) $28 - 11 = 17$ (exact)
$4.2 \times 5.30 \times 17 = 378.42$ (calc), 3.8×10^2 (corr)
c) 28 (exact) $- 4.2 = 23.8$ (calc and corr)
5.30×11 (exact) $= 58.3$ (calc and corr)
$\dfrac{23.8}{58.3} = 0.40823328$ (calc), 0.408 (corr)
d) 28 (exact) $- 4.2 = 23.8$ (calc and corr)
11 (exact) $- 5.30 = 5.7$ (calc), 5.70 (corr)
$\dfrac{23.8}{5.70} = 4.1754386$ (calc), 4.18 (corr)

2.139 a) 2.07×10^2, 243, 1.03×10^3 b) 2.11×10^{-3}, 0.0023, 3.04×10^{-2}
c) $23{,}000$, 9.67×10^4, 2.30×10^5 d) 0.000014, 0.00013, 1.5×10^{-4}

Answers to Multiple-Choice Practice Test

MC 2.1	d	MC 2.2	b	MC 2.3	d	MC 2.4	b	MC 2.5	b
MC 2.6	c	MC 2.7	a	MC 2.8	e	MC 2.9	a	MC 2.10	c
MC 2.11	c	MC 2.12	b	MC 2.13	c	MC 2.14	e	MC 2.15	e
MC 2.16	c	MC 2.17	c	MC 2.18	d	MC 2.19	b	MC 2.20	d

CHAPTER THREE
Unit Systems and Dimensional Analysis

METRIC SYSTEM UNITS (SECS. 3.1–3.4)

3.1 a) volume   b) mass   c) length   d) mass

3.3 a) Nano   b) Mega   c) Milli   d) Deci

3.5 a) μg   b) dm   c) cL   d) pg

3.7

	Metric Prefix	Abbreviation for Prefix	Mathematical Meaning of Prefix
	milli-	m	10^{-3}
a)	tera-	T	10^{12}
b)	nano-	n	10^{-9}
c)	giga-	G	10^{9}
d)	micro-	μ	10^{-6}

3.9

	Metric Unit	Property Being Measured	Abbreviation for Metric Unit
	microliter	volume	μL
a)	kilogram	mass	kg
b)	megameter	length	Mm
c)	nanogram	mass	ng
d)	milliliter	volume	mL

3.11 a) 6.8 dm   b) 3.2 pL   c) 7.23 cL   d) 6.5 Mg

3.13 a) length   b) area   c) volume   d) volume

3.15 a) milligram   b) centimeter   c) millimeter   d) liter

3.17 a) 1 inch   b) 1 meter   c) 1 pound   d) 1 gallon

AREA AND VOLUME MEASUREMENTS (SECS. 3.4 AND 3.5)

3.19 a) $A = (4.52 \text{ cm})^2 = 20.4304$ (calc) $= 20.4 \text{ cm}^2$ (corr)
b) $A = (3.5 \text{ m}) \times (9.2 \text{ m}) = 32.2$ (calc) $= 32 \text{ m}^2$ (corr)

c) $A = 3.142 \times (4.579 \text{ mm})^2 = 65.879071 \text{ (calc)} = 65.88 \text{ mm}^2 \text{ (corr)}$

d) $A = 1/2 \times (5.5 \text{ mm}) \times (3.0 \text{ mm}) = 8.25 \text{ (calc)} = 8.2 \text{ mm}^2 \text{ (corr)}$

3.21   a) $V = (5.4 \text{ cm}) \times (0.52 \text{ cm}) \times (3.4 \text{ cm}) = 9.5472 \text{ (calc)} = 9.5 \text{ cm}^3 \text{ (corr)}$

b) $V = 3.142 \times (2.4 \text{ cm})^2 \times (7.5 \text{ cm}) = 135.7344 \text{ (calc)} = 1.4 \times 10^2 \text{ cm}^3 \text{ (corr)}$

c) $V = 4/3 \times 3.142 \times (87 \text{ mm})^3 = 2.7586886 \times 10^6 \text{ (calc)} = 2.8 \times 10^6 \text{ mm}^3 \text{ (corr)}$

d) $V = (7.2 \text{ cm})^3 = 373.248 \text{ (calc)} = 3.7 \times 10^2 \text{ cm}^3$

3.23   a) equal   b) not equal   c) not equal   d) equal

CONVERSION FACTORS (SEC. 3.6)

3.25   a) $24 \text{ hr} = 1 \text{ day}$ $\dfrac{1 \text{ day}}{24 \text{ hr}}$ $\dfrac{24 \text{ hr}}{1 \text{ day}}$ b) $60 \text{ sec} = 1 \text{ min}$ $\dfrac{1 \text{ min}}{60 \text{ sec}}$ $\dfrac{60 \text{ sec}}{1 \text{ min}}$

c) $10 \text{ decades} = 1 \text{ century}$ $\dfrac{1 \text{ century}}{10 \text{ decades}}$ $\dfrac{10 \text{ decades}}{1 \text{ century}}$

d) $365 \text{ days} = 1 \text{ yr}$ $\dfrac{1 \text{ yr}}{365 \text{ days}}$ $\dfrac{365 \text{ days}}{1 \text{ yr}}$

3.27   a) 1 quart = 2 pints, 1 pint = 2 cups   b) 1 yard = 3 feet
            c) 1 ton = 2000 pounds   d) 1 yard = 36 inches

3.29   a) $\dfrac{1 \text{ kL}}{10^3 \text{ L}}$ $\dfrac{10^3 \text{ L}}{1 \text{ kL}}$ b) $\dfrac{1 \text{ mg}}{10^{-3} \text{ g}}$ $\dfrac{10^{-3} \text{ g}}{1 \text{ mg}}$ c) $\dfrac{1 \text{ cm}}{10^{-2} \text{ m}}$ $\dfrac{10^{-2} \text{ m}}{1 \text{ cm}}$ d) $\dfrac{1 \text{ } \mu\text{sec}}{10^{-6} \text{ sec}}$ $\dfrac{10^{-6} \text{ sec}}{1 \text{ } \mu\text{sec}}$

3.31   a) $A = 10^3 \text{ g}$ b) $A = 10^{-6} \text{ m}$ c) $A = 453.6 \text{ g}$ d) $A = 2.540 \text{ cm}$

3.33   a) 4 significant figures   b) exact   c) 4 significant figures   d) exact

3.35   a) exact   b) inexact   c) exact   d) exact

DIMENSIONAL ANALYSIS—METRIC–METRIC UNIT CONVERSIONS (SEC. 3.7)

3.37   a) $37 \text{ L} \times \dfrac{1 \text{ dL}}{10^{-1} \text{ L}} = 3.7 \times 10^2 \text{ dL}$ (calc and corr)

b) $37.0 \text{ mm} \times \dfrac{10^{-3} \text{ m}}{1 \text{ mm}} = 3.70 \times 10^{-2} \text{ m}$ (calc and corr)

c) $0.37 \text{ pg} \times \dfrac{10^{-12} \text{ g}}{1 \text{ pg}} = 3.7 \times 10^{-13} \text{ g}$ (calc and corr)

d) $370 \text{ kL} \times \dfrac{10^3 \text{ L}}{1 \text{ kL}} = 3.7 \times 10^5 \text{ L}$ (calc and corr)

3.39   a) $47 \text{ Mg} \times \dfrac{10^6 \text{ g}}{1 \text{ Mg}} \times \dfrac{1 \text{ mg}}{10^{-3} \text{ g}} = 4.7 \times 10^{10} \text{ mg}$ (calc and corr)

b) $5.00 \text{ nL} \times \dfrac{10^{-9} \text{ L}}{1 \text{ nL}} \times \dfrac{1 \text{ cL}}{10^{-2} \text{ L}} = 5.00 \times 10^{-7} \text{ cL}$ (calc and corr)

c) $6 \times 10^{-2} \text{ } \mu\text{m} \times \dfrac{10^{-6} \text{ m}}{1 \text{ } \mu\text{m}} \times \dfrac{1 \text{ dm}}{10^{-1} \text{ m}} = 6 \times 10^{-7} \text{ dm}$ (calc and corr)

d) $37 \text{ pm} \times \dfrac{10^{-12} \text{ m}}{1 \text{ pm}} \times \dfrac{1 \text{ km}}{10^3 \text{ m}} = 3.7 \times 10^{-14} \text{ km}$ (calc and corr)

3.41 a) $365 \text{ m}^2 \times \left(\dfrac{1 \text{ km}}{10^3 \text{ m}}\right)^2 = 3.65 \times 10^{-4} \text{ km}^2$ (calc and corr)

b) $365 \text{ m}^2 \times \left(\dfrac{1 \text{ cm}}{10^{-2} \text{ m}}\right)^2 = 3.65 \times 10^6 \text{ cm}^2$ (calc and corr)

c) $365 \text{ m}^2 \times \left(\dfrac{1 \text{ dm}}{10^{-1} \text{ m}}\right)^2 = 3.65 \times 10^4 \text{ dm}^2$ (calc and corr)

d) $365 \text{ m}^2 \times \left(\dfrac{1 \text{ Mm}}{10^6 \text{ m}}\right)^2 = 3.65 \times 10^{-10} \text{ Mm}^2$ (calc and corr)

3.43 a) $35 \text{ m}^3 \times \left(\dfrac{1 \text{ mm}}{10^{-3} \text{ m}}\right)^3 = 3.5 \times 10^{10} \text{ mm}^3$ (calc and corr)

b) $35 \text{ m}^3 \times \left(\dfrac{1 \text{ pm}}{10^{-12} \text{ m}}\right)^3 = 3.5 \times 10^{37} \text{ pm}^3$ (calc and corr)

c) $35 \text{ m}^3 \times \left(\dfrac{1 \text{ Gm}}{10^9 \text{ m}}\right)^3 = 3.5 \times 10^{-26} \text{ Gm}^3$ (calc and corr)

d) $35 \text{ m}^3 \times \left(\dfrac{1 \text{ } \mu\text{m}}{10^{-6} \text{ m}}\right)^3 = 3.5 \times 10^{19} \text{ } \mu\text{m}^3$ (calc and corr)

3.45 a) $67 \text{ cm}^3 \times \left(\dfrac{10^{-2} \text{ m}}{1 \text{ cm}}\right)^3 \times \left(\dfrac{1 \text{ km}}{10^3 \text{ m}}\right)^3 = 6.7 \times 10^{-14} \text{ km}^3$ (calc and corr)

b) $67 \text{ cm}^3 \times \left(\dfrac{10^{-2} \text{ m}}{1 \text{ cm}}\right)^3 \times \left(\dfrac{1 \text{ mm}}{10^{-3} \text{ m}}\right)^3 = 6.7 \times 10^4 \text{ mm}^3$ (calc and corr)

c) $67 \text{ cm}^3 \times \left(\dfrac{10^{-2} \text{ m}}{1 \text{ cm}}\right)^3 \times \left(\dfrac{1 \text{ dm}}{10^{-1} \text{ m}}\right)^3 = 6.7 \times 10^{-2} \text{ dm}^3$ (calc and corr)

d) $67 \text{ cm}^3 \times \left(\dfrac{10^{-2} \text{ m}}{1 \text{ cm}}\right)^3 \times \left(\dfrac{1 \text{ Mm}}{10^6 \text{ m}}\right)^3 = 6.7 \times 10^{-23} \text{ Mm}^3$ (calc and corr)

DIMENSIONAL ANALYSIS—METRIC–ENGLISH UNIT CONVERSIONS (SEC. 3.7)

3.47 a) $100.0 \text{ yd} \times \dfrac{3 \text{ ft}}{1 \text{ yd}} \times \dfrac{12 \text{ in.}}{1 \text{ ft}} \times \dfrac{2.540 \text{ cm}}{1 \text{ in.}} \times \dfrac{10^{-2} \text{ m}}{1 \text{ cm}} = 91.44 \text{ m}$ (calc and corr)

b) $100.0 \text{ yd} \times \dfrac{3 \text{ ft}}{1 \text{ yd}} \times \dfrac{12 \text{ in.}}{1 \text{ ft}} \times \dfrac{2.540 \text{ cm}}{1 \text{ in.}} = 9144 \text{ cm}$ (calc and corr)

c) $100.0 \text{ yd} \times \dfrac{3 \text{ ft}}{1 \text{ yd}} \times \dfrac{12 \text{ in.}}{1 \text{ ft}} \times \dfrac{2.540 \text{ cm}}{1 \text{ in.}} \times \dfrac{10^{-2} \text{ m}}{1 \text{ cm}} \times \dfrac{1 \text{ km}}{10^3 \text{ m}} = 0.09144 \text{ km}$ (calc and corr)

d) $100.0 \text{ yd} \times \dfrac{3 \text{ ft}}{1 \text{ yd}} \times \dfrac{12 \text{ in.}}{1 \text{ ft}} = 3600$ (calc) $= 3.600 \times 10^3 \text{ in.}$ (corr)

3.49 a) $75 \text{ mL} \times \dfrac{10^{-3} \text{ L}}{1 \text{ mL}} \times \dfrac{1 \text{ qt}}{0.9463 \text{ L}} = 0.079256$ (calc) $= 0.079 \text{ qt}$ (corr)

b) $75 \text{ mL} \times \dfrac{10^{-3} \text{ L}}{1 \text{ mL}} \times \dfrac{1 \text{ qt}}{0.9463 \text{ L}} \times \dfrac{1 \text{ gal}}{4 \text{ qt}} = 0.019814$ (calc) $= 0.020 \text{ gal}$ (corr)

c) $75 \text{ mL} \times \dfrac{10^{-3} \text{ L}}{1 \text{ mL}} \times \dfrac{1 \text{ qt}}{0.9463 \text{ L}} \times \dfrac{32 \text{ fl oz}}{1 \text{ qt}} = 2.5361936$ (calc) $= 2.5 \text{ fl oz}$ (corr)

d) $75 \text{ mL} \times \dfrac{1 \text{ cm}^3}{1 \text{ mL}} = 75 \text{ cm}^3$ (calc and corr)

3.51   a) 6.6×10^{21} tons $\times \dfrac{2000 \text{ lb}}{1 \text{ ton}} \times \dfrac{453.6 \text{ g}}{1 \text{ lb}} = 5.98752 \times 10^{27}$ (calc) $= 6.0 \times 10^{27}$ g (corr)

     b) 6.6×10^{21} tons $\times \dfrac{2000 \text{ lb}}{1 \text{ ton}} \times \dfrac{453.6 \text{ g}}{1 \text{ lb}} \times \dfrac{1 \text{ kg}}{10^3 \text{ g}} = 5.98752 \times 10^{24}$ (calc) $= 6.0 \times 10^{24}$ kg (corr)

     c) 6.6×10^{21} tons $\times \dfrac{2000 \text{ lb}}{1 \text{ ton}} \times \dfrac{453.6 \text{ g}}{1 \text{ lb}} \times \dfrac{1 \text{ ng}}{10^{-9} \text{ g}} = 5.98752 \times 10^{36}$ (calc) $= 6.0 \times 10^{36}$ ng (corr)

     d) 6.6×10^{21} tons $\times \dfrac{2000 \text{ lb}}{1 \text{ ton}} \times \dfrac{16 \text{ oz}}{1 \text{ lb}} = 2.112000 \times 10^{26}$ (calc) $= 2.1 \times 10^{26}$ oz (corr)

3.53   a) $61 \text{ cm}^3 \times \left(\dfrac{1 \text{ in.}}{2.540 \text{ cm}}\right)^3 \times \left(\dfrac{1 \text{ ft}}{12 \text{ in.}}\right)^3 = 2.1541947 \times 10^{-3}$ (calc) $= 2.2 \times 10^{-3}$ ft^3 (corr)

     b) $61 \text{ cm}^3 \times \left(\dfrac{1 \text{ in.}}{2.540 \text{ cm}}\right)^3 \times \left(\dfrac{1 \text{ ft}}{12 \text{ in.}}\right)^3 \times \left(\dfrac{1 \text{ yd}}{3 \text{ ft}}\right)^3 = 7.978507 \times 10^{-5}$ (calc)

$$= 8.0 \times 10^{-5} \text{ yd}^3 \text{ (corr)}$$

     c) $61 \text{ cm}^3 \times \left(\dfrac{1 \text{ in.}}{2.540 \text{ cm}}\right)^3 = 3.7224484$ (calc) $= 3.7$ in.^{3} (corr)

     d) $61 \text{ cm}^3 \times \left(\dfrac{1 \text{ in.}}{2.540 \text{ cm}}\right)^3 \times \left(\dfrac{1 \text{ ft}}{12 \text{ in.}}\right)^3 \times \left(\dfrac{1 \text{ mi}}{5280 \text{ ft}}\right)^3 = 1.463468 \times 10^{-14}$ (calc)

$$= 1.5 \times 10^{-14} \text{ mi}^3 \text{ (corr)}$$

3.55   a) $A = 2.1 \text{ cm} \times 2.5 \text{ cm} = 5.25$ (calc) $= 5.2$ cm^2 (corr)

     b) $A = 2.1 \text{ cm} \times 2.5 \text{ cm} \times \left(\dfrac{1 \text{ in.}}{2.540 \text{ cm}}\right)^2 = 0.81375163$ (calc) $= 0.81$ in.^{2} (corr)

3.57   a) $95 \text{ cm} \times 105 \text{ cm} \times 145 \text{ cm} = 1.446375 \times 10^6$ cm^3 (calc)

$$= 1.4 \times 10^6 \text{ cm}^3 \text{ (corr)}$$

     b) $(95 \times 105 \times 145) \text{ cm}^3 \times \left(\dfrac{1 \text{ in.}}{2.540 \text{ cm}}\right)^3 \times \left(\dfrac{1 \text{ ft}}{12 \text{ in.}}\right)^3 = 51.07825108$ ft^3 (calc) $= 51$ ft^3 (corr)

DIMENSIONAL ANALYSIS—UNITS INVOLVING TWO TYPES OF MEASUREMENTS (SEC. 3.7)

3.59   a) $\dfrac{55 \text{ L}}{1 \text{ sec}} \times \dfrac{60 \text{ sec}}{1 \text{ min}} \times \dfrac{60 \text{ min}}{1 \text{ hr}} = 198,000$ (calc) $= 2.0 \times 10^5$ L/hr (corr)

     b) $\dfrac{55 \text{ L}}{1 \text{ sec}} \times \dfrac{1 \text{ kL}}{10^3 \text{ L}} = 5.5 \times 10^{-2}$ kL/sec (calc and corr)

     c) $\dfrac{55 \text{ L}}{1 \text{ sec}} \times \dfrac{1 \text{ dL}}{10^{-1} \text{ L}} \times \dfrac{60 \text{ sec}}{\text{min}} = 33,000$ (calc) $= 3.3 \times 10^4$ dL/min (corr)

     d) $\dfrac{55 \text{ L}}{1 \text{ sec}} \times \dfrac{1 \text{ mL}}{10^{-3} \text{ L}} \times \dfrac{60 \text{ sec}}{1 \text{ min}} \times \dfrac{60 \text{ min}}{1 \text{ hr}} \times \dfrac{24 \text{ hr}}{1 \text{ day}} = 4.752 \times 10^9$ (calc) $= 4.8 \times 10^9$ mL/day (corr)

3.61   a) $\dfrac{0.0057 \text{ }\mu\text{g}}{1 \text{ mL}} \times \dfrac{1 \text{ mL}}{10^{-3} \text{ L}} = 5.7 \text{ }\mu\text{g/L}$ (calc and corr)

     b) $\dfrac{0.0057 \text{ }\mu\text{g}}{1 \text{ mL}} \times \dfrac{10^{-6} \text{ g}}{1 \text{ }\mu\text{g}} = 5.7 \times 10^{-9}$ g/mL (calc and corr)

c) $\dfrac{0.0057\ \mu g}{1\ mL} \times \dfrac{1\ mL}{10^{-3}\ L} \times \dfrac{10^{-6}\ L}{1\ \mu L} \times \dfrac{10^{-6}\ g}{1\ \mu g} \times \dfrac{1\ mg}{10^{-3}\ g} = 5.7 \times 10^{-9}\ mg/\mu L$ (calc and corr)

d) $\dfrac{0.0057\ \mu g}{1\ mL} \times \dfrac{1\ mL}{10^{-3}\ L} \times \dfrac{10^{3}\ L}{1\ kL} \times \dfrac{10^{-6}\ g}{1\ \mu g} \times \dfrac{1\ kg}{10^{3}\ g} = 5.7 \times 10^{-6}\ kg/kL$ (calc and corr)

DENSITY (SEC. 3.8)

3.63   a) $\dfrac{10.0\ g}{6.37\ mL} = 1.569858713$ (calc) $= 1.57\ g/mL$ (corr)

b) $\dfrac{12.0\ g}{4.0\ mL} = 3$ (calc) $= 3.0\ g/mL$ (corr)

c) $\dfrac{6.23\ g}{4.02\ cm^3} \times \dfrac{1\ cm^3}{1\ mL} = 1.549751244$ (calc) $= 1.55\ g/mL$ (corr)

d) $\dfrac{0.111\ lb}{9.5\ mL} \times \dfrac{453.6\ g}{1\ lb} = 5.299957895$ (calc) $= 5.3\ g/mL$ (corr)

3.65   Use density as a conversion factor.

a) $47.6\ mL \times \dfrac{0.791\ g}{1\ mL} = 37.6516$ (calc) $= 37.7\ g$ (corr)

b) $47.6\ cm^3 \times \dfrac{10.40\ g}{1\ cm^3} = 495.04$ (calc) $= 495\ g$ (corr)

c) $47.6\ L \times \dfrac{1.25\ g}{1\ L} = 59.5\ g$ (calc and corr)

d) $47.6\ cm^3 \times \dfrac{2.18\ g}{1\ cm^3} = 103.768$ (calc) $= 104\ g$ (corr)

3.67   Use density as a conversion factor.

a) $17.6\ g \times \dfrac{1\ mL}{1.027\ g} = 17.13729309$ (calc) $= 17.1\ mL$ (corr)

b) $17.6\ g \times \dfrac{1\ cm^3}{19.3\ g} \times \dfrac{1\ mL}{1\ cm^3} = 0.9119170984$ (calc) $= 0.912\ mL$ (corr)

c) $17.6\ g \times \dfrac{1\ L}{1.29\ g} \times \dfrac{1\ mL}{10^{-3}\ L} = 13643.41085$ (calc) $= 1.36 \times 10^{4}\ mL$ (corr)

d) $17.6\ g \times \dfrac{1\ cm^3}{1.008\ g} \times \dfrac{1\ mL}{1\ cm^3} = 17.460317$ (calc) $= 17.5\ mL$ (corr)

3.69 $(5.261 - 3.006)\ g = 2.255\ g$ red liquid (calc and corr) $d = \dfrac{2.255\ g}{2.171\ mL} = 1.0386918$ (calc)

$= 1.039\ g/mL$ (corr)

3.71   pathway: $gal \rightarrow qt \rightarrow L \rightarrow mL \rightarrow g \rightarrow lb$

$13.0\ gal \times \dfrac{4\ qt}{1\ gal} \times \dfrac{0.9463\ L}{1\ qt} \times \dfrac{1\ mL}{10^{-3}\ L} \times \dfrac{0.56\ g}{1\ mL} \times \dfrac{1\ lb}{453.6\ g} = 60.750123$ (calc) $= 61\ lb$ (corr)

3.73   Al = aluminum, Cr = chromium

$$100.0 \text{ g Al} \times \frac{1 \text{ cm}^3 \text{ Al}}{2.70 \text{ g Al}} \times \frac{1 \text{ cm}^3 \text{ Cr}}{1 \text{ cm}^3 \text{ Al}} \times \frac{7.18 \text{ g Cr}}{1 \text{ cm}^3 \text{ Cr}} = 265.92593 \text{ (calc)} = 266 \text{ g (corr)}$$

3.75   a) float, less dense than water   b) sink, more dense than water

EQUIVALENCE CONVERSION FACTORS (SEC. 3.9)

3.77   a) equivalence   b) equality   c) equality   d) equivalence

3.79   a) $15.9 \text{ kg} \times \dfrac{32 \text{ mg}}{1 \text{ kg}} = 508.8 \text{ (calc)} = 510 \text{ mg (corr)}$

        b) $3750 \text{ mg} \times \dfrac{1 \text{ kg}}{32 \text{ mg}} = 117.1875 \text{ (calc)} = 120 \text{ kg (corr)}$

3.81   a) $2.30 \text{ L} \times \dfrac{2.30 \text{ }\mu\text{g}}{1 \text{ L}} \times \dfrac{10^{-6} \text{ g}}{1 \text{ }\mu\text{g}} = 5.29 \times 10^{-6} \text{ g (calc and corr)}$

        b) $2.30 \text{ mL} \times \dfrac{10^{-3} \text{ L}}{1 \text{ mL}} \times \dfrac{2.30 \text{ }\mu\text{g}}{1 \text{ L}} \times \dfrac{10^{-6} \text{ g}}{1 \text{ }\mu\text{g}} = 5.29 \times 10^{-9} \text{ g (calc and corr)}$

        c) $2.30 \text{ qt} \times \dfrac{0.9463 \text{ L}}{1 \text{ qt}} \times \dfrac{2.30 \text{ }\mu\text{g}}{1 \text{ L}} \times \dfrac{10^{-6} \text{ g}}{1 \text{ }\mu\text{g}} = 5.005927 \times 10^{-6} \text{ g (calc)} = 5.01 \times 10^{-6} \text{ g (corr)}$

        d) $2.30 \text{ fl oz} \times \dfrac{1 \text{ qt}}{32 \text{ fl oz}} \times \dfrac{0.9463 \text{ L}}{1 \text{ qt}} \times \dfrac{2.30 \text{ }\mu\text{g}}{1 \text{ L}} \times \dfrac{10^{-6} \text{ g}}{1 \text{ }\mu\text{g}} = 0.156435218 \times 10^{-6} \text{ (calc)}$

$$= 1.56 \times 10^{-7} \text{ g (corr)}$$

3.83   a) $16.0 \text{ gal} \times \dfrac{1 \text{ sec}}{0.20 \text{ gal}} \times \dfrac{1 \text{ min}}{60 \text{ sec}} = \text{(calc)} = 1.33333333 \text{ min (calc)} = 1.3 \text{ min (corr)}$

        b) $13.0 \text{ qt} \times \dfrac{1 \text{ gal}}{4 \text{ qt}} \times \dfrac{1 \text{ sec}}{0.20 \text{ gal}} \times \dfrac{1 \text{ min}}{60 \text{ sec}} = 0.270833333 \text{ min (calc)} = 0.27 \text{ min (corr)}$

        c) $7.00 \text{ L} \times \dfrac{1 \text{ qt}}{0.9463 \text{ L}} \times \dfrac{1 \text{ gal}}{4 \text{ qt}} \times \dfrac{1 \text{ sec}}{0.20 \text{ gal}} \times \dfrac{1 \text{ min}}{60 \text{ sec}} = 0.154108986 \text{ min (calc)} = 0.15 \text{ min (corr)}$

        d) $5355 \text{ mL} \times \dfrac{10^{-3} \text{ L}}{1 \text{ mL}} \times \dfrac{1 \text{ qt}}{0.9463 \text{ L}} \times \dfrac{1 \text{ gal}}{4 \text{ qt}} \times \dfrac{1 \text{ sec}}{0.20 \text{ gal}} \times \dfrac{1 \text{ min}}{60 \text{ sec}} = 0.117893374 \text{ (calc)}$

$$= 0.12 \text{ min (corr)}$$

3.85   a) $\$1000.00 \times \dfrac{1 \text{ bushel}}{\$9.20} = 108.6956522 \text{ (calc)} = 109 \text{ bushels (corr)}$

        b) $45 \text{ bushels} \times \dfrac{\$9.20}{1 \text{ bushel}} = \$414.00 \text{ (calc and corr)}$

PERCENTAGE AND PERCENT ERROR (SEC. 3.10)

3.87   a) $\% \text{ copper} = \dfrac{2.902 \text{ g copper}}{3.053 \text{ g penny}} \times 100 = 95.054045 \text{ (calc)} = 95.05\% \text{ copper (corr)}$

        b) mass zinc = 3.053 g penny − 2.902 g copper = 0.151 g zinc

           $\% \text{ zinc} = \dfrac{0.151 \text{ g zinc}}{3.053 \text{ g penny}} \times 100 = 4.945948 \text{ (calc)} = 4.95\% \text{ zinc (corr)}$

           Alternate method: % zinc = 100 − % copper = 100 − 95.05 = 4.95% zinc (calc and corr)

3.89 $65.3 \text{ g mixture} \times \dfrac{34.2 \text{ g water}}{100 \text{ g mixture}} = 22.3326 \text{ (calc)} = 22.3 \text{ g water (corr)}$

3.91 $437 \text{ g solution} \times \dfrac{15.3 \text{ g salt}}{100 \text{ g solution}} = 66.861 \text{ (calc)} = 66.9 \text{ g salt (corr)}$

3.93 $1.00 \text{ gal solution} \times \dfrac{4 \text{ qt}}{1 \text{ gal}} \times \dfrac{0.9463 \text{ L}}{1 \text{ qt}} \times \dfrac{1 \text{ mL}}{10^{-3} \text{ L}} \times \dfrac{1.013 \text{ g}}{1 \text{ mL}} = 3834.4076 \text{ g of solution}$

$3834.4076 \text{ g solution} \times \dfrac{9.00 \text{ g acetic acid}}{100 \text{ g solution}} = 345.09668 \text{ (calc)} = 345 \text{ g acetic acid (corr)}$

3.95 $661 \text{ Gummi} \times \dfrac{30.9 \text{ bears}}{100 \text{ Gummi}} \times \dfrac{23.0 \text{ orange}}{100 \text{ bears}} \times \dfrac{6.4 \text{ one-ear}}{100 \text{ orange}} = 3.0065453 \text{ (calc)} = 3 \text{ one-ear (corr)}$

3.97   student 1: $\dfrac{(78.0 - 78.5)°\text{C}}{78.5°\text{C}} \times 100 = -0.63694267\% \text{ (calc)} = -0.6\% \text{ (corr)}$

student 2: $\dfrac{(77.9 - 78.5)°\text{C}}{78.5°\text{C}} \times 100 = -0.76433121\% \text{ (calc)} = -0.8\% \text{ (corr)}$

student 3: $\dfrac{(79.7 - 78.5)°\text{C}}{78.5°\text{C}} \times 100 = 1.5286624\% \text{ (calc)} = 1.5\% \text{ (corr)}$

TEMPERATURE SCALES (SEC. 3.11)

3.99   a) $32°\text{F}$ b) $0°\text{C}$ c) 273 K

3.101 $20°\text{C} \times \dfrac{9°\text{F}}{5°\text{C}} = 36°\text{F} \text{ (calc and corr)}$

3.103   a) $1352°\text{C} = 1352°\text{C}$ above FP $\times \dfrac{9°\text{F}}{5°\text{C}} = 2433.6 \text{ (calc)}$

$= 2434 \text{ (corr) }°\text{F}$ above FP $(32°\text{F}) = 2434 + 32 = 2466°\text{F (corr)}$

b) $37.6°\text{C} = 37.6°\text{C}$ above FP $\times \dfrac{9°\text{F}}{5°\text{C}} = 67.68 \text{ (calc)}$

$= 67.7 \text{ (corr) }°\text{F}$ above FP $(32°\text{F}) = 67.7 + 32 = 99.7°\text{F (corr)}$

c) $1352°\text{F} = 1352 - 32 = 1320°\text{F}$ above FP $\times \dfrac{5°\text{C}}{9°\text{F}}$

$= 733.3333 \text{ (calc)} = 733 \text{ (corr) }°\text{C}$ above FP $(0°\text{C}) = 733°\text{C}$

d) $37.6°\text{F} = 37.6 - 32 = 5.6°\text{F}$ above FP $\times \dfrac{5°\text{F}}{9°\text{F}}$

$= 3.11111 \text{ (calc)} = 3.1 \text{ (corr) }°\text{C}$ above FP $(0°\text{C}) = 3.1°\text{C}$

3.105   a) $\text{K} = 275°\text{C} + 273 = 548 \text{ K (calc and corr)}$
b) $\text{K} = 275.2°\text{C} + 273.2 = 548.4 \text{ K (calc and corr)}$
c) $\text{K} = 275.73°\text{C} + 275.15 = 548.88 \text{ K (calc and corr)}$
d) $°\text{C} = 275 \text{ K} - 273 = 2°\text{C (calc and corr)}$

3.107   a) $101°\text{F} = 101 - 32 = 69°\text{F}$ above FP $\times \dfrac{5°\text{C}}{9°\text{F}} = 38.33333 \text{ (calc)}$

$= 38 \text{ (corr) }°\text{C}$ above FP $(0°\text{C}) = 38°\text{C}$

b) $-218.4°C = 218.4°C$ below FP $\times \dfrac{9°F}{5°C} = 393.12$ (calc) $= 393.1$ (corr) °F below FP

$$(32°F) = (32 - 393.1) = -361.1°F$$

c) $804°C + 273 = 1077$ K (calc and corr)

d) 77 K $- 273 = -196°C$ (calc and corr) $-196°C = 196°C$ below FP $\times \dfrac{9°F}{5°C}$

$$= 352.8 \text{ (calc)} = 353 \text{ (corr) °F below FP } (32°F) = (32 - 353) = -321°F$$

3.109   a) convert $223°C$ to K and compare
   K $= 223°C + 273 = 496$ K, which is higher than 223 K

b) convert $100°C$ to °F and compare

$100°C$ above FP $\times \dfrac{9°F}{5°C} = 180°F + 32 = 212°F$,

which is higher than $100°F$

c) convert $-15°C$ to °F and compare

$15°C$ below FP $\times \dfrac{9°F}{5°C} = 27°F$ below FP $(32°F) = 32 - 27 = 5°F$,

which is higher than $4°F$

d) convert $-40°C$ to °F and compare

$40°C$ below FP $\times \dfrac{9°F}{5°C} = 72°F$ below FP $(32°F) = 32 - 72 = -40°F$,

which is the same as $-40°F$

Multi-Concept Problems

3.111   a) 4.3 cm $\times \dfrac{10^{-2}\text{ m}}{1\text{ cm}} = 4.3 \times 10^{-2}$ m (calc and corr), 2 significant figures

b) 4.03 cm $\times \dfrac{10^{-2}\text{ m}}{1\text{ cm}} = 4.03 \times 10^{-2}$ m (calc and corr), 3 significant figures

c) 0.33030 cm $\times \dfrac{10^{-2}\text{ m}}{1\text{ cm}} = 3.3030 \times 10^{-3}$ m (calc and corr), 5 significant figures

d) 5.123 cm $\times \dfrac{10^{-2}\text{ m}}{1\text{ cm}} = 5.123 \times 10^{-2}$ m (calc and corr), 4 significant figures

3.113   Before adding, all values must have the same units; convert them all to centimeters.

(1) 20.9 dm $\times \dfrac{10^{-1}\text{ m}}{1\text{ dm}} \times \dfrac{1\text{ cm}}{10^{-2}\text{ m}} = 209$ cm   (2) 2030 mm $\times \dfrac{10^{-3}\text{ m}}{1\text{ mm}} \times \dfrac{1\text{ cm}}{10^{-2}\text{ m}} = 203$ cm

(3) 1.90 m $\times \dfrac{1\text{ cm}}{10^{-2}\text{ m}} = 190$ cm   (4) 0.00183 km $\times \dfrac{10^{3}\text{ m}}{1\text{ km}} \times \dfrac{1\text{ cm}}{10^{-2}\text{ m}} = 183$ cm

(5) 203 cm $= 203$ cm 209 cm $+ 203$ cm $+ 190$ cm $+ 183$ cm $+ 203$ cm $= 988$ cm

Average $= \dfrac{988\text{ cm}}{5} = 197.6$ (calc) $= 198$ cm (corr)

3.115   a) 3.72 mm $\times \dfrac{10^{-3}\text{ m}}{1\text{ mm}} = 3.72 \times 10^{-3}$ m (calc and corr)

b) 3.720 mm $\times \dfrac{10^{-3}\text{ m}}{1\text{ mm}} = 3.720 \times 10^{-3}$ m (calc and corr)

   c) $0.0372 \text{ mm} \times \dfrac{10^{-3} \text{ m}}{1 \text{ mm}} = 3.72 \times 10^{-5} \text{ m (calc and corr)}$

   d) $0.037200 \text{ mm} \times \dfrac{10^{-1} \text{ L}}{1 \text{ dL}} = 3.7200 \times 10^{-5} \text{ m (calc and corr)}$

3.117 $\quad 8 \text{ yr} \times \dfrac{365 \text{ days}}{1 \text{ yr}} \times \dfrac{24 \text{ hr}}{1 \text{ day}} \times \dfrac{60 \text{ min}}{1 \text{ day}} \times \dfrac{69 \text{ beats}}{1 \text{ min}} = 2.90131200 \times 10^8 \text{ beats (calc)}$

$$= 2.9 \times 10^8 \text{ beats (corr)}$$

3.119 $\quad$ pathway: $\text{pt} \rightarrow \text{qt} \rightarrow \text{L} \rightarrow \text{mL} \rightarrow \text{g}$

$\quad 1.00 \text{ pt} \times \dfrac{1 \text{ qt}}{2 \text{ pt}} \times \dfrac{0.9463 \text{ L}}{1 \text{ qt}} \times \dfrac{1 \text{ mL}}{10^{-3} \text{ L}} \times \dfrac{1.05 \text{ g}}{1 \text{ mL}} = 496.8075 \text{ (calc)} = 497 \text{ g (corr)}$

3.121 $\quad 325 \text{ g acid} \times \dfrac{100 \text{ g solution}}{38.1 \text{ g acid}} \times \dfrac{1 \text{ cm}^3 \text{ solution}}{1.29 \text{ g solution}} \times \dfrac{1 \text{ mL solution}}{1 \text{ cm}^3 \text{ solution}} = 661.25455 \text{ (calc)} = 661 \text{ mL (corr)}$

3.123 $\quad$ a) pathway $\quad \begin{array}{l} \nearrow \ \mu\text{g} \rightarrow \text{g} \rightarrow \text{mg} \\ \searrow \ \text{mL} \rightarrow \text{L} \rightarrow \text{dL} \end{array}$

$\quad 5000 \dfrac{1 \ \mu\text{g}}{1 \text{ mL}} \times \dfrac{10^{-6} \text{ g}}{1 \ \mu\text{g}} \times \dfrac{1 \text{ mg}}{10^{-3} \text{ g}} \times \dfrac{1 \text{ mL}}{10^{-3} \text{ L}} \times \dfrac{10^{-1} \text{ L}}{1 \text{ dL}} = 500 \dfrac{\text{mg}}{\text{dL}} \text{ (calc and corr)}$

$\qquad\qquad\qquad\qquad\qquad\qquad\qquad\qquad\qquad\qquad\qquad \therefore$ Yes, a life-threatening situation.

   b) pathway $\quad \begin{array}{l} \nearrow \ \text{g} \rightarrow \text{mg} \\ \searrow \ \text{L} \rightarrow \text{dL} \end{array}$

$\quad 0.5 \dfrac{1 \text{ g}}{1 \text{ L}} \times \dfrac{1 \text{ mg}}{10^{-3} \text{ g}} \times \dfrac{10^{-1} \text{ L}}{1 \text{ dL}} = 50 \dfrac{\text{mg}}{\text{dL}} \text{ (calc and corr)} \quad \therefore$ No, not a life-threatening situation.

3.125 $\quad \text{Area (cm}^2) = 4.0 \text{ in.} \times 4.0 \text{ in.} \times \left(\dfrac{2.540 \text{ cm}}{1 \text{ in.}}\right)^2 = 103.2256 \text{ cm}^2 \text{ (calc)} = 1.0 \times 10^2 \text{ cm}^2 \text{ (corr)}$

$\quad \text{Volume (cm}^3) = 0.466 \text{ g} \times \left(\dfrac{1 \text{ cg}}{10^{-2} \text{ g}} \times \dfrac{1 \text{ cm}^3}{269 \text{ cg}}\right) = 0.1732342 \text{ (calc)} = 0.173 \text{ cm}^3 \text{ (corr)}$

$\quad \text{Thickness} = \dfrac{0.173 \text{ cm}^3}{1.0 \times 10^2 \text{ cm}^2} = 0.00173 \text{ cm} = 0.0173 \text{ mm (calc)} = 1.7 \times 10^{-2} \text{ mm (corr)}$

3.127 $\quad$ Conversion factor: $(212 - 32)°\text{F} = [\,200 - (-200)\,]°\text{H}; \ 180°\text{F} = 400°\text{H}; \ 20°\text{H} = 9°\text{F}$

$\quad 50°\text{F} = 50 - 32 = 18°\text{F above FP} \times \dfrac{20°\text{H}}{9°\text{F}} = 40°\text{H above FP} \ (-200°\text{H}) = 40 + (-200)$

$$= -160°\text{H}$$

Answers to Multiple-Choice Practice Test

MC 3.1	b	**MC 3.2**	e	**MC 3.3**	e	**MC 3.4**	c	**MC 3.5**	c	**MC 3.6**	a
MC 3.7	c	**MC 3.8**	a	**MC 3.9**	c	**MC 3.10**	c	**MC 3.11**	d	**MC 3.12**	b
MC 3.13	b	**MC 3.14**	b	**MC 3.15**	b	**MC 3.16**	d	**MC 3.17**	c	**MC 3.18**	e
MC 3.19	e	**MC 3.20**	c								

CHAPTER FOUR
Basic Concepts about Matter

PHYSICAL STATES OF MATTER (SEC. 4.2)

4.1   a) shape (indefinite vs. definite)   b) indefinite shape

4.3   a) Does not take shape of the container, definite volume.
b) Takes on shape of container, indefinite volume.
c) Does not take shape of the container, definite volume.
d) Takes on shape of container, definite volume.

4.5   a) solid   b) liquid   c) liquid   d) gas

PROPERTIES OF MATTER (SEC. 4.3)

4.7   a) physical   b) chemical   c) physical   d) chemical

4.9   a) chemical   b) physical   c) physical   d) physical

4.11   a) extensive   b) intensive   c) intensive   d) intensive

4.13   a) intensive   b) intensive   c) intensive   d) intensive

CHANGES IN MATTER (SEC. 4.4)

4.15   a) physical   b) chemical   c) physical   d) physical

4.17   a) chemical   b) physical   c) chemical   d) physical

4.19   a) physical   b) physical   c) chemical   d) chemical

4.21   a) physical   b) physical   c) chemical   d) physical

4.23   a) freezing   b) condensation   c) sublimation   d) evaporation

PURE SUBSTANCES AND MIXTURES (SECS. 4.5 AND 4.6)

4.25   a) heterogeneous and homogeneous mixtures   b) pure substance, homogeneous mixture

4.27   a) false   b) true   c) false   d) true

4.29   a) heterogeneous mixture   b) homogeneous mixture
c) homogeneous mixture   d) heterogeneous mixture

4.31   a) homogeneous mixture, 1 phase   b) homogeneous mixture, 1 phase
c) heterogeneous mixture, 2 phases   d) heterogeneous mixture, 2 phases

4.33   a) chemically homogeneous, physically homogeneous
           b) chemically heterogeneous, physically homogeneous
           c) chemically heterogeneous, physically heterogeneous
           d) chemically heterogeneous, physically heterogeneous

ELEMENTS AND COMPOUNDS (SEC. 4.7)

4.35   a) compound   b) compound
           c) no classification possible d) no classification possible

4.37   a) true   b) false   c) false   d) false

4.39   a) A (no classification possible), B (no classification possible), C (compound)
           b) D (compound), E (no classification possible), F (no classification possible),
              G (no classification possible)

4.41   First box, mixture; second box, compound

4.43   a) applies   b) applies   c) applies   d) does not apply

DISCOVERY AND ABUNDANCE OF THE ELEMENTS (SEC. 4.8)

4.45   a) false   b) true   c) false   d) true

4.47   a) true   b) true   c) false   d) true

4.49   a) Yes, silicon is more abundant.   b) No, hydrogen is more abundant.
           c) No, oxygen is more abundant.   d) Yes, sodium is more abundant.

NAMES AND CHEMICAL SYMBOLS OF THE ELEMENTS (SEC. 4.9)

4.51   B, boron   C, carbon   F, fluorine   H, hydrogen
           I, iodine   K, potassium   N, nitrogen   O, oxygen
           P, phosphorus S, sulfur   U, uranium   V, vanadium
           W, tungsten   Y, yttrium

4.53   a) silver, Ag   b) iron, Fe   c) mercury, Hg   d) potassium, K

4.55   a) Ar designates argon   b) Ca designates calcium
           c) Ti designates titanium   d) Ra designates radium

4.57   a) incorrect   b) incorrect   c) incorrect   d) incorrect

4.59   a) potassium, K   b) hydrogen, H   c) tungsten, W   d) iodine, I

4.61   oxygen, O   silicon, Si   aluminum, Al   hydrogen, H
           calcium, Ca   magnesium, Mg

4.63   a) Rebecca   Re-Be-C-Ca
           b) Nancy   Na-N-C-Y
           c) Sharon   S-H-Ar-O-N
           d) Alice   Al-I-Ce

Multi-Concept Problems

4.65 a) colorless gas, odorless gas, colorless liquid, boils at 43°C
b) toxic to humans, Ni reacts with CO

4.67 a) compound   b) mixture   c) compound   d) mixture

4.69 a) homogeneous mixture   b) element
c) heterogeneous mixture   d) homogeneous mixture

4.71 a) heterogeneous mixture   b) heterogeneous mixture
c) homogeneous mixture   d) heterogeneous, but not a mixture

4.73 (b) and (c)

4.75 Alphabetized: Bh, Bi, Bk, Cf, Co, Cs, Cu, Hf, Ho, Hs, In, Nb, Ni, No, Np, Os, Pb, Po, Pu, Sb, Sc, Si, Sn, Yb

Answers to Multiple-Choice Practice Test

MC 4.1	e	MC 4.2	c	MC 4.3	c	MC 4.4	d	MC 4.5	d	MC 4.6	c
MC 4.7	e	MC 4.8	a	MC 4.9	d	MC 4.10	a	MC 4.11	d	MC 4.12	e
MC 4.13	b	MC 4.14	d	MC 4.15	e	MC 4.16	d	MC 4.17	e	MC 4.18	c
MC 4.19	a	MC 4.20	b								

CHAPTER FIVE
Atoms, Molecules, and Subatomic Particles

ATOMS AND MOLECULES (SECS. 5.1 AND 5.2)

5.1 a) consistent   b) consistent   c) not consistent   d) consistent

5.3 a) heteroatomic   b) heteroatomic   c) homoatomic   d) heteroatomic

5.5 a) tetratomic   b) tetratomic   c) triatomic   d) hexatomic

5.7 a) compound   b) compound   c) element   d) compound

5.9 a) False, molecules must contain two or more atoms.
b) true
c) False, some compounds have molecules as their basic structural unit and others have ions as their basic structural unit.
d) False, the diameter is approximately 10^{-10} meters.

5.11 a) one substance, homoatomic—element
b) two substances, molecules are homoatomic—mixture
c) one phase present, two kinds of triatomic heteroatomic molecules—mixture
d) two kinds of molecules—mixture

CHEMICAL FORMULAS (SEC. 5.4)

5.13 a) AB_2 b) BC_2 c) B_2C_2 d) A_3B_2C

5.15 a) compound   b) compound   c) element   d) element

5.17 a) $C_{20}H_{30}$ b) H_2SO_4 c) HCN   d) $KMnO_4$

5.19 a) NH_4NO_3 − 9 total atoms   b) K_3PO_4 − 8 total atoms
c) $Al(ClO)_3$ − 7 total atoms   d) $Ca(ClO_4)_2$ − 11 total atoms

5.21 a) NH_4NO_3 − 3 oxygen atoms   b) K_3PO_4 − 4 oxygen atoms
c) $AlPO_4$ − 4 oxygen atoms   d) $Ca(ClO_4)_2$ − 8 oxygen atoms

5.23 a) NO − 1 atom nitrogen, 1 atom oxygen, No − 1 atom nobelium
b) Cs_2 − 2 atoms cesium, CS_2 − 1 atom carbon, 2 atoms sulfur
c) $CoBr_2$ − 1 atom cobalt, 2 atoms bromine, $COBr_2$ − 1 atom carbon, 1 atom oxygen, 2 atoms bromine
d) H − 1 atom hydrogen, H_2 − 2 atoms hydrogen

5.25   a) H_3PO_4 b) $SiCl_4$ c) NO_2 d) H_2O_2

5.27   a) HCN   b) H_2SO_4

SUBATOMIC PARTICLES (SEC. 5.5)

5.29   a) electron   b) proton   c) proton, neutron   d) neutron

5.31   a) false   b) false   c) false   d) true

5.33   Relative mass of proton to electron

$$= \frac{\text{mass of proton}}{\text{mass of electron}} = \frac{1.673 \times 10^{-24}\,\text{g}}{9.109 \times 10^{-28}\,\text{g}} = 1836.6450763\ (\text{calc}) = 1837\ (\text{corr})$$

ATOMIC NUMBER AND MASS NUMBER (SEC. 5.6)

5.35   a) calcium, atomic number is 20   b) carbon, atomic number is 6
         c) sulfur, atomic number is 16   d) copper, atomic number is 29

5.37   a) atomic number is 24, mass number is 53   b) atomic number is 44, mass number is 103
         c) atomic number is 101, mass number is 256   d) atomic number is 16, mass number is 34

5.39   a) 20 protons + 22 neutrons = 42 (mass number)
         b) 20 protons + 20 neutrons = 40 (mass number)
         c) 20 protons + 19 neutrons = 39 (mass number)
         d) 20 protons + 21 neutrons = 41 (mass number)

5.41   a) (mass number) 70 = 34 protons + 36 neutrons
         b) (mass number) 70 = 35 protons + 35 neutrons
         c) (mass number) 70 = 33 protons + 37 neutrons
         d) (mass number) 70 = 36 protons + 34 neutrons

5.43   atomic number = number of protons   a) 5   b) 8   c) 13   d) 18

5.45   mass number = number of protons + number of neutrons
         a) 5 + 6 = 11   b) 8 + 8 = 16   c) 13 + 14 = 27   d) 18 + 22 = 40

5.47   number of nucleons = number of protons + number of neutrons
         a) 5 + 6 = 11   b) 8 + 8 = 16   c) 13 + 14 = 27   d) 18 + 22 = 40

5.49   charge on nucleus = number of protons
         a) +5   b) +8   c) +13   d) +18

5.51   a) $^{11}_{5}B$ b) $^{16}_{8}O$ c) $^{27}_{13}Al$ d) $^{40}_{18}Ar$

5.53   a) atomic number   b) atomic number and mass number
         c) mass number   d) atomic number and mass number

5.55   a) Nitrogen, N   b) Aluminum, Al   c) Barium, Ba   d) Gold, Au

5.57

	Symbol	Atomic Number	Mass Number	Number of Protons	Number of Neutrons
	$_2^3$He	2	3	2	1
a)	$_{28}^{60}$Ni	28	60	28	32
b)	$_{18}^{37}$Ar	18	37	18	19
c)	$_{38}^{90}$Sr	38	90	38	52
d)	$_{92}^{235}$U	92	235	92	143

5.59   a) 27 protons, 27 electrons, 32 neutrons   b) 45 protons, 45 electrons, 58 neutrons
　　　　　　c) 69 protons, 69 electrons, 100 neutrons   d) 9 protons, 9 electrons, 10 neutrons

5.61   a) total subatomic particles = 11 protons + 11 electrons + 12 neutrons = 34
　　　　　　b) total subatomic particles in the nucleus = 11 protons + 12 neutrons = 23
　　　　　　c) total number nucleons = total subatomic particles in the nucleus = 11 protons + 12 neutrons = 23
　　　　　　d) total charge in the nucleus = +11, +1 for each proton, neutrons have no charge

5.63   a) atom has 15 protons; mass number = 15 + 16;　$_{15}^{31}$P
　　　　　　b) oxygen has 8 protons; mass number is 8 + 10 = 18;　$_8^{18}$O
　　　　　　c) chromium has 24 protons;　$_{24}^{54}$Cr
　　　　　　d) A gold atom has 79 protons and 79 electrons.
　　　　　　　　276 subatomic particles − 79 protons − 79 electrons = 118 neutrons.
　　　　　　　　∴ mass number = 79 + 118 = 197;　$_{79}^{197}$Au

5.65   a) 12 protons + 12 electrons = 24 subatomic particles that carry a charge
　　　　　　b) 37 subatomic particles − 24 subatomic particles that carry a charge (protons + electrons)
　　　　　　　　　　　　　　　　　　　　　　　　= 13 subatomic particles with no charge (neutrons)
　　　　　　c) 12 protons + 13 neutrons = 25 particles in the nucleus
　　　　　　d) 37 subatomic particles − 25 particles in the nucleus
　　　　　　　　　　　　　　　　　　　　　　　　= 12 particles not found in the nucleus (electrons)

5.67   a) same total number of subatomic particles, 60
　　　　　　b) same number of neutrons, 16
　　　　　　c) same number of neutrons, 12
　　　　　　d) same number of electrons, 3

ISOTOPES (SEC. 5.7)

5.69   a) 35 protons + 45 neutrons = 80 (mass number), symbol $_{35}^{80}$Br
　　　　　　b) 6 protons + 8 neutrons = 14 (mass number), symbol $_6^{14}$C
　　　　　　c) 11 protons + 12 neutrons = 23 (mass number), symbol $_{11}^{23}$Na
　　　　　　d) 50 protons + 70 neutrons = 120 (mass number), symbol $_{50}^{120}$Sn

5.71 $_{40}^{96}$Zr　$_{40}^{94}$Zr　$_{40}^{92}$Zr　$_{40}^{91}$Zr　$_{40}^{90}$Zr

5.73   a) C, 13.003 amu, mass number = 13   b) Ar, 35.968 amu, mass number = 36
　　　　　　c) Co, 58.933 amu, mass number = 59   d) Pd, 103.904 amu, mass number = 104

5.75   a) To determine the mass numbers, we round the amu to the nearest whole number.

53.940 amu = mass number of 54, thus our symbol is ^{54}Fe

55.935 amu = mass number of 56, thus our symbol is ^{56}Fe

56.935 amu = mass number of 57, thus our symbol is ^{57}Fe

57.933 amu = mass number of 58, thus our symbol is ^{58}Fe

b) The percent abundance for the third isotope can be calculated from the percent abundance of the other three isotopes.

First isotope   =   5.82%

Second isotope   = 91.66%

Fourth isotope   =   0.33%

Total   = 97.81%   By definition, the percent abundance must = 100; thus, the percent abundance of the third isotope = 100% − 97.81% = 2.19%

5.77   a) not isotopes, different atomic numbers
          b) not isotopes, different atomic numbers
          c) not isotopes, different atomic numbers

5.79   a) no   b) yes   c) no   d) yes

5.81   a) isotopes   b) isobars   c) neither   d) isotopes

5.83   a) isotopes   b) isobars   c) neither   d) isotopes

5.85 $^{40}_{18}Ar$ $^{40}_{19}K$ $^{40}_{20}Ca$

ATOMIC MASSES (SEC. 5.8)

5.87   a) nitrogen, 14.001 amu   b) gold, 196.97 amu
          c) uranium, 238.03 amu   d) iodine, 126.90 amu

5.89   a) atomic number 37, rubidium   b) atomic number 26, iron
          c) atomic number 41, niobium   d) atomic number 9, fluorine

5.91   Values on this hypothetical relative mass scale are $Q = 8.00$ bebs, $X = 4.00$ bebs, and $Z = 2.00$ bebs.

5.93   a) $Z = \frac{3}{4}(12) = 9$ amu; $X = 3Z = 27$ amu; $Q = 9(12) = 108$ amu
          b) Z is Be, X is Al, Q is Ag

5.95   0.220 × 271 lb   = 59.62 (calc)   =   59.6 lb (corr)
          0.190 × 175 lb   = 33.25 (calc)   =   33.2 lb (corr)
          0.260 × 263 lb   = 68.38 (calc)   =   68.4 lb (corr)
          0.150 × 182 lb   = 27.3 (calc)   =   27.3 lb (corr)
          0.190 × 191 lb   = 36.29 (calc)   =   36.3 lb (corr)
                                                        = 224.8 lb (calc and corr)

5.97   0.7553 × 34.9689 amu = 26.41201 (calc)   = 26.41 amu (corr)
          0.2447 × 36.9659 amu = 9.0455557 (calc)   =   9.046 amu (corr)
                                                        = 35.456 amu (calc)
                                                        = 35.46 amu (corr)

5.99

0.0793×45.95263 amu	$= 3.6440435$ (calc)	$= 3.64$ amu (corr)
0.0728×46.95176 amu	$= 3.418088128$ (calc)	$= 3.42$ amu (corr)
0.7394×47.94795 amu	$= 35.452714$ (calc)	$= 35.45$ amu (corr)
0.0551×48.94787 amu	$= 2.6970276$ (calc)	$= 2.70$ amu (corr)
0.0534×49.94479 amu	$= 2.667051786$ (calc)	$= \underline{2.67 \text{ amu}}$ (corr)

$= 47.88$ amu (calc and corr)

5.101 a) ^{14}N b) ^{51}V c) ^{121}Sb d) ^{193}Ir

5.103 a) possible   b) possible   c) possible   d) not possible

5.105

	Isotope A	Isotope B	Atomic Mass	Abundance % A	Abundance % B
	46 amu	48 amu	47 amu	50%	50%
a)	46 amu	50 amu	47 amu	75%	25%
b)	46 amu	50 amu	49 amu	25%	75%
c)	45 amu	50 amu	46 amu	80%	20%
d)	45 amu	50 amu	49 amu	20%	80%

5.107 a) $(K, 39.10 \text{ amu})X = Sc, 44.96 \text{ amu}$

$$X = \frac{Sc, 44.96 \text{ amu}}{K, 39.10 \text{ amu}} = 1.149872123 \text{ (calc)} = 1.15 \text{ (corr)}$$

b) $(Ca, 40.08 \text{ amu})X = Zn, 65.38 \text{ amu}$

$$X = \frac{Zn, 65.38 \text{ amu}}{Ca, 40.08 \text{ amu}} = 1.631237525 \text{ (calc)} = 1.63 \text{ (corr)}$$

c) $(O, 16.00 \text{ amu})X = P, 30.97 \text{ amu}$

$$X = \frac{P, 30.97 \text{ amu}}{O, 16.00 \text{ amu}} = 1.935625 \text{ (calc)} = 1.94 \text{ (corr)}$$

d) $(B, 10.81 \text{ amu})X = Ne, 20.18 \text{ amu}$

$$X = \frac{Ne, 20.18 \text{ amu}}{B, 10.81 \text{ amu}} = 1.866790009 \text{ (calc)} = 1.87 \text{ (corr)}$$

5.109 The number 12.0000 amu applies only to the ^{12}C isotope. The number 12.01 amu applies to naturally occurring carbon, a mixture of ^{12}C, ^{13}C, and ^{14}C.

EVIDENCE SUPPORTING THE EXISTENCE AND ARRANGEMENT OF SUBATOMIC PARTICLES (SEC. 5.9)

5.111 a) true   b) false   c) false   d) false

Multi-Concept Problems

5.113 a) $X = 7$ total atoms $- (2 \text{ Na atoms} + 3 \text{ O atoms}) = 2$ $Na_2S_2O_3$
       b) $2X = 9$ total atoms $- (1 \text{ Ba atom} + 2 \text{ Cl atoms}) = 6, X = 3$ $Ba(ClO_3)_2$
       c) $2X = 16$ total atoms $- (10 \text{ O atoms}) = 6, X = 3$ $Na_3P_3O_{10}$
       d) $4X = 24$ total atoms, $X = 6$ $C_6H_{12}O_6$

5.115 a) heteroatomic molecules, N_2O, H_2O, CCl_4, CH_2Br_2
         b) pentaatomic molecules, CCl_4, CH_2Br_2
         c) different compounds, N_2O, H_2O, CCl_4, CH_2Br_2 (O_2 is an element)
         d) total atoms, given 3 molecules of each component

$$O_2, 2 \text{ atoms} \times 3 = 6 \text{ total atoms}$$
$$N_2O, 3 \text{ atoms} \times 3 = 9 \text{ total atoms}$$
$$H_2O, 3 \text{ atoms} \times 3 = 9 \text{ total atoms}$$
$$CCl_4, 5 \text{ atoms} \times 3 = 15 \text{ total atoms}$$
$$CH_2Br_2, 5 \text{ atoms} \times 3 = \underline{15 \text{ total atoms}}$$
$$54 \text{ total atoms}$$

5.117 a) false   b) false   c) true   d) true

5.119 a) $^{37}_{18}Ar$, $^{39}_{19}K$, $^{42}_{20}Ca$, $^{44}_{21}Sc$, $^{43}_{22}Ti$ b) $^{44}_{21}Sc$, $^{42}_{20}Ca$, $^{43}_{22}Ti$, $^{39}_{19}K$, $^{37}_{18}Ar$
         c) $^{37}_{18}Ar$, $^{39}_{19}K$, $^{42}_{20}Ca$, $^{44}_{21}Sc$, $^{43}_{22}Ti$ d) $^{44}_{21}Sc$, $^{43}_{22}Ti$, $^{42}_{20}Ca$, $^{39}_{19}K$, $^{37}_{18}Ar$

5.121 a) $^{8}_{5}B$ b) $^{12}_{5}B$ c) $^{12}_{5}B$ d) $^{16}_{5}B$

5.123 a) 29 and 29   b) 29 and 29   c) 34 and 36

5.125 a) 13   b) 27   c) 14   d) 27.0 amu

5.127 a) $1.679 \times 30.97 \text{ amu} = 51.99863 \text{ (calc)} = 52.00 \text{ amu} = Cr$
         b) $30e \times 3 = 90e = Th$
         c) $4p + 20p = 24p = Cr$
         d) $[20e + 20p = 40 \text{ e\&p}] \times 2 = 80 \text{ e\&p} = 40p = Zr$

5.129 Nickel weighs 20.000 amu on the new scale and 58.6934 amu on the old carbon-12 scale. Therefore, 20.000 new amu units is equal to 58.6934 old amu units.

a) $107.87 \text{ old amu} \times \dfrac{20.000 \text{ new amu}}{58.6934 \text{ old amu}} = 36.757114 \text{ (calc)} = 36.757 \text{ new amu (corr)}$

b) $196.97 \text{ old amu} \times \dfrac{20.000 \text{ new amu}}{58.6934 \text{ old amu}} = 67.118279 \text{ (calc)} = 67.118 \text{ new amu (corr)}$

Answers to Multiple-Choice Practice Test

MC 5.1 d   **MC 5.2** c   **MC 5.3** a   **MC 5.4** b   **MC 5.5** d   **MC 5.6** c

MC 5.7 e   **MC 5.8** a   **MC 5.9** e   **MC 5.10** d   **MC 5.11** e   **MC 5.12** c

MC 5.13 c   **MC 5.14** d   **MC 5.15** e   **MC 5.16** c   **MC 5.17** a   **MC 5.18** c

MC 5.19 e   **MC 5.20** c

CHAPTER SIX
Electronic Structure and Chemical Periodicity

PERIODIC LAW AND PERIODIC TABLE (SECS. 6.1 and 6.2)

6.1 a) $_3$Li is in the second row of elements, period 2, and is in the first column on the left side of the periodic table, group IA.

b) $_{16}$S is in the third row of elements, period 3, and is in the third column from the right side of the periodic table, group VIA.

c) $_{47}$Ag is in the fifth row of elements, period 5, and is in the eighth column from the right side of the periodic table, group IB.

d) $_{83}$Bi is in the sixth row of elements, period 6, and is in the fourth column from the right side of the periodic table, group VA.

6.3 a) Al   b) Be   c) Sn   d) K

6.5 a) $_{19}$K and $_{37}$Rb   b) $_{15}$P and $_{33}$As   c) $_9$F and $_{53}$I   d) $_{11}$Na and $_{55}$Cs

6.7 a) group   b) periodic law   c) periodic law   d) period

6.9 a) bromine   b) lithium   c) argon   d) strontium

6.11 a) F   b) Be   c) Ne   d) Li

6.13 a) 3   b) 1   c) 2   d) 2

TERMINOLOGY ASSOCIATED WITH ELECTRON ARRANGEMENTS (SECS. 6.3–6.6)

6.15 a) 2   b) 10   c) 6   d) 14

6.17 Any orbital has a maximum capacity of 2 electrons.
a) 2   b) 2   c) 2   d) 2

6.19 a) 1   b) 1   c) 5   d) 3

6.21 a) fifth electron shell, $n = 5$, contains 4 subshells (s, p, d, f)
b) the subshells in the fifth electron shell with the lowest energy, s and p
c) fifth electron shell, can hold a maximum of 32 electrons
d) fifth electron shell, contains 16 electron orbitals

6.23 a) true   b) false   c) true   d) true   e) false

6.25 a) spherical   b) dumbbell   c) cloverleaf   d) spherical

6.27   a) allowed
           b) allowed
           c) not allowed, $2p$ is lowest allowed
           d) not allowed, $4f$ is lowest allowed

ELECTRON CONFIGURATIONS (SEC. 6.7)

6.29   a) $3p$ b) $5s$ c) $4d$ d) $4p$

6.31   a) $3s$ b) $3p$ c) $4f$ d) $3d$

6.33   a) $1s, 2s, 3s, 4s$ b) $4s, 4p, 4d, 4f$
           c) $2p, 6s, 4f, 5d$ d) $3p, 4s, 3d, 5p$

6.35   a) $_{13}$Al, $1s^2\,2s^2\,2p^6\,3s^2\,3p^1$
           b) $_{26}$Fe, $1s^2\,2s^2\,2p^6\,3s^2\,3p^6\,4s^2\,3d^6$
           c) $_{53}$I, $1s^2\,2s^2\,2p^6\,3s^2\,3p^6\,4s^2\,3d^{10}\,4p^6\,5s^2\,4d^{10}\,5p^5$
           d) $_{88}$Ra, $1s^2\,2s^2\,2p^6\,3s^2\,3p^6\,4s^2\,3d^{10}\,4p^6\,5s^2\,4d^{10}\,5p^6\,6s^2\,4f^{14}\,5d^{10}\,6p^6\,7s^2$

6.37   a) $_{10}$Ne   b) $_{19}$K   c) $_{22}$Ti   d) $_{30}$Zn

6.39   a) 2 electrons in electron shell 1, $1s^2$
              8 electrons in electron shell 2, $2s^2\,2p^6$
              5 electrons in electron shell 3, $3s^2\,3p^3$ chemical symbol, P
           b) 6 total "s" electrons
           c) 9 total "p" electrons
           d) 0 "d" electrons

6.41   a) $[\,\text{Ne}\,]\,3s^2 3p^1$ b) $[\,\text{Ar}\,]\,4s^1$
           c) $[\,\text{Ar}\,]\,4s^2 3d^{10}$ d) $[\,\text{Kr}\,]\,5s^2 4d^{10} 5p^2$

6.43   a) $[\,\text{He}\,]\,2s^2 2p^4$ b) $[\,\text{Ne}\,]\,3s^2$
           c) $[\,\text{Ar}\,]\,4s^2$ d) $[\,\text{Ar}\,]\,4s^2 3d^{10} 4p^5$

6.45   a) magnesium   b) chlorine   c) tellurium   d) barium

6.47   a) 10 core electrons   b) 10 core electrons
           c) 36 core electrons   d) 54 core electrons

6.49   a) 2 outer electrons   b) 7 outer electrons
           c) 16 outer electrons   d) 2 outer electrons

ORBITAL DIAGRAMS (SEC. 6.8)

6.51   a) ↑↓ ↑
              $1s$ $2s$
           b) ↑↓ ↑↓ ↑↓ ↑↓ ↑
              $1s$ $2s$ $2p$
           c) ↑↓ ↑↓ ↑↓ ↑↓ ↑↓ ↑↓ ↑ ↑ ↑
              $1s$ $2s$ $2p$ $3s$ $3p$
           d) ↑↓ ↑↓ ↑↓ ↑↓ ↑↓ ↑↓ ↑↓ ↑↓ ↑↓ ↑↓ ↑↓ ↑↓ ↑↓ ↑
              $1s$ $2s$ $2p$ $3s$ $3p$ $4s$ $3d$

6.53 a) ↑↓   ↑↓   ↑ ↑
       1s   2s   2p

      b) ↑↓   ↑↓   ↑↓ ↑↓ ↑↓
       1s   2s   2p

      c) ↑↓   ↑↓   ↑↓ ↑↓ ↑↓   ↑
       1s   2s   2p   3s

      d) ↑↓   ↑↓   ↑↓ ↑↓ ↑↓   ↑↓   ↑ ↑ ↑
       1s   2s   2p   3s   3p

6.55 a) chlorine, $[\text{Ne}]$ ↓↑   ↓↑ ↓↑ ↑
                      $3s^2$ $3p^5$

      b) manganese, $[\text{Ar}]$ ↓↑   ↑ ↑ ↑ ↑ ↑
                        $4s^2$ $3d^5$

      c) strontium, $[\text{Kr}]$ ↓↑
                      $5s^2$

      d) aluminum, $[\text{Ne}]$ ↓↑   ↑
                    $3s^2$ $3p^1$

6.57 a) Mg, $3s^2$ ↑↓
              3s

      b) Cl, $3p^5$ ↑↓ ↑↓ ↑
               3p

      c) Ca, $4s^2$ ↑↓
               4s

      d) Co, $3d^7$ ↑↓ ↑↓ ↑ ↑ ↑
               3d

6.59 a) Li, 1 unpaired electron   b) N, 3 unpaired electrons
      c) S, 2 unpaired electrons   d) Ga, 1 unpaired electron

6.61 a) boron, paramagnetic   b) nitrogen, paramagnetic
      c) fluorine, paramagnetic   d) sulfur, paramagnetic

ELECTRON CONFIGURATIONS AND THE PERIODIC LAW (SEC. 6.9)

6.63 a) no   b) yes   c) no   d) yes

ELECTRON CONFIGURATIONS AND THE PERIODIC TABLE (SEC. 6.10)

6.65 a) p area   b) d area   c) d area   d) s area

6.67 a) p subshell   b) d subshell   c) d subshell   d) s subshell

6.69 a) 3p subshell   b) 3d subshell   c) 4d subshell   d) 6s subshell

6.71 a) 3, Cr, Ru, and Pt   b) 3, S, Xe, and element 114
      c) 2, Ca, and element 114   d) 2, Ca, and Ru

6.73 a) Al   b) Li   c) La   d) Sc

6.75 a) Kr   b) Li   c) K   d) Lu

6.77   a) $1s^22s^22p^63s^23p^64s^23d^{10}4p^3$

          b) $1s^22s^22p^63s^23p^64s^23d^{10}4p^65s^24d^{10}5p^3$

          c) $1s^22s^22p^63s^23p^64s^23d^{10}4p^65s^1$

          d) $1s^22s^22p^63s^23p^64s^23d^{10}4p^65s^24d^{10}$

6.79   a) $1s^22s^22p^63s^2$ b) $1s^22s^22p^63s^23p^64s^2$

          c) $1s^22s^22p^63s^23p^64s^23d^{10}$ d) $1s^22s^22p^63s^23p^64s^23d^{10}4p^65s^24d^{10}5p^2$

6.81   a) Ti, $1s^22s^22p^63s^23p^64s^23d^2$ ∴ 2 electrons in $3d$

          b) Ni, $1s^22s^22p^63s^23p^64s^23d^8$ ∴ 8 electrons in $3d$

          c) Se, $1s^22s^22p^63s^23p^64s^23d^{10}4p^4$ ∴ 10 electrons in $3d$

          d) Pd, $1s^22s^22p^63s^23p^64s^23d^{10}4p^65s^24d^8$ ∴ 10 electrons in $3d$

6.83   a) 4   b) 9   c) 1   d) 0

CLASSIFICATION SYSTEMS FOR THE ELEMENTS (SEC. 6.11)

6.85   a) metallic   b) nonmetallic   c) metallic   d) nonmetallic

6.87   Period 2 elements: Li and Be are metals; B, C, N, O, F and Ne are nonmetals.

6.89   Group VA elements: N, P and As are nonmetals; Sb and Bi are metals.

6.91   a) period 4 metal, K   b) period 2 nonmetal, B

          c) group IIA metal, Be   d) group VA nonmetal, N

6.93   a) 1 is a metal (Rb).   b) 3 is a nonmetal (S).

          c) 2 is a good conductor of electricity (Al).   d) 4 is a good conductor of heat (Sn).

6.95   a) transition elements, $_{30}$Zn, $_{74}$W

          b) noble gases, $_{36}$Kr

          c) representative elements, $_{34}$Se, $_{36}$K, $_{38}$Sr

          d) inner transition elements, $_{58}$Ce

6.97   a) lowest-atomic-numbered nonmetal, H

          b) lowest-atomic-numbered transition metal, Sc

          c) lowest-atomic-numbered period 4 representative metal, K

          d) lowest-atomic-numbered period 4 representative nonmetal, As

6.99   a) transition elements—$3d, 4d, 5d, 6d$ = 4 periods of transition elements in the current periodic table

          b) representative elements, s and p subshells, $s^1, s^2, p^1, p^2, p^3, p^4, p^5$ = 7 groups

          c) periods that contain metals, Periods 2, 3, 4, 5, 6, 7 = 6 periods

          d) no periods consist of only metals; all periods contain at least one noble gas

6.101   a) 1 is a noble gas element (Xe).

          b) 1 is an inner transition element (U).

          c) 3 are nonmetal elements (H, S, and Xe).

          d) 2 are metallic representative elements (element 114 and Ca).

CHEMICAL PERIODICITY (SEC. 6.12)

6.103   a) false   b) true   c) false

6.105   Period 2 elements: Li and Be are metals; B is a metalloid; and C, N, O, F, and Ne are nonmetals.

6.107   a) $_{37}$Rb   b) $_{22}$Ti   c) $_{34}$Se   d) $_{56}$Ba

6.109   a) $_9$F   b) $_{15}$P   c) $_{30}$Zn   d) $_{17}$Cl

6.111   a) Bi is the most metallic in Group VA.
        b) Beryllium is the most nonmetallic in Group IIA.
        c) Potassium is the most metallic in Period 4.
        d) Neon is the most nonmetallic in Period 2.

6.113   a) B   b) Ga   c) K   d) Rb

6.115   a) Nitrogen has the smallest atomic radius in Group VA.
        b) Radium has the largest atomic radius in Group IIA.
        c) Krypton has the smallest atomic radius in Period 4.
        d) Lithium has the largest atomic radius in Period 2.

Multi-Concept Problems

6.117   a) The $2s$ subshell is more than filled. It has a maximum occupancy of 2 electrons.
        b) The $2p$ subshell must be filled before the $3s$. It can accommodate 6 electrons.
        c) The $2p$ subshell, which fills after the $2s$, is omitted.
        d) The $3p$ and $4s$ subshells fill after the $3s$ and before the $3d$ subshell.

6.119   a) Period 3, Group IA   b) Period 3, Group IIIA
        c) Period 4, Group IIIB   d) Period 4, Group VIIA

6.121   a) 1 unpaired electron   b) 1 unpaired electron
        c) 1 unpaired electron   d) 1 unpaired electron

6.123   a) paramagnetic   b) paramagnetic
        c) paramagnetic   d) paramagnetic

6.125   a) above: $[\text{Ne}]\,3s^2$, below: $[\text{Kr}]\,5s^2$
        b) above: $[\text{Ar}]\,4s^2 3d^3$, below: $[\text{Xe}]\,6s^2 4f^{14} 5d^3$
        c) above: $[\text{He}]\,2s^2 2p^5$, below: $[\text{Ar}]\,4s^2 3d^{10} 4p^5$
        d) above: $[\text{Ar}]\,4s^2 3d^{10} 4p^1$, below: $[\text{Xe}]\,6s^2 4f^{14} 5d^{10} 6p^1$

6.127   a) $1s^2 2s^2 2p^6 3s^2 3p^2$ (Si)   b) $1s^2 2s^2 2p^2$ (C)
        c) $1s^2 2s^2 2p^3$ (N)   d) $1s^2 2s^2 2p^3$ (N)

6.129   a) Be   b) Be   c) Ne   d) Ar

6.131   a) Groups IVA and VIA   b) Group VB and middle column of Group VIIIB
        c) Group IVB and last column of Group VIIIB   d) Group IA

6.133   a) Po   b) Cr   c) Elements 88–116, 118   d) Elements 12–18

6.135   a) F   b) Ag

6.137 a) heaviest alkaline earth metal, Ra
           b) lightest metalloid in Group VA, As
           c) halogen in period 6, At
           d) lightest representative element, H

Answers to Multiple-Choice Practice Test

MC 6.1   b   MC 6.2   e   MC 6.3   a   MC 6.4   c   MC 6.5   e   MC 6.6   d

MC 6.7   c   MC 6.8   c   MC 6.9   c   MC 6.10 b   MC 6.11 b   MC 6.12 c

MC 6.13 a   MC 6.14 d   MC 6.15 d   MC 6.16 e   MC 6.17 c   MC 6.18 b

MC 6.19 c   MC 6.20 c

CHAPTER SEVEN
Chemical Bonds

VALENCE ELECTRONS (SEC. 7.2)

7.1   Valence electrons are all of the s and p electrons in the highest shell.
a) $2 + 3 = 5$ b) 2 c) 2 d) $2 + 4 = 6$

7.3   a) $2 + 2 = 4$ b) $2 + 1 = 3$ c) $2 + 4 = 6$ d) 2

7.5   Valence electrons for a representative element are all the "s" and "p" electrons in the outermost shell.
a) Lithium, $1s^2 2s^1$ has one valence electron in its $2s$ subshell.
b) Nitrogen, $1s^2 2s^2 2p^3$ has five valence electrons in its $2s$ and $2p$ subshells.
c) Chlorine, $1s^2 2s^2 2p^6 3s^2 3p^6$ has seven valence electrons in its $3s$ and $3p$ subshells.
d) Aluminum, $1s^2 2s^2 2p^6 3s^2 3p^1$ has three valence electrons in its $3s$ and $3p$ subshells.

7.7   a) 1 element has 5 valence electrons (Bi).
b) 1 element has 7 valence electrons (I).
c) 2 elements have 2 valence electrons (Mg and Ba).
d) 1 element has 8 valence electrons (Ne).

7.9   a) Group IA, 1 valence electron
b) Group VIIIA, 8 valence electrons
c) Group IIA, 2 valence electrons
d) Group VIIA, 7 valence electrons

7.11   a) C   b) F   c) Mg   d) P

LEWIS SYMBOLS FOR ATOMS (SEC. 7.2)

7.13   a) $\cdot$ C $\cdot$ b) $\cdot$ Si $\cdot$ c) $\cdot$ Cl :   d) $\cdot$ Ba $\cdot$

7.15   a) B   b) C   c) F   d) Li

7.17   a) Incorrect, O has 6 valence electrons.   b) Correct, Al has 3 valence electrons.
c) Incorrect, Cl has 7 valence electrons.   d) Correct, Se has 6 valence electrons.

7.19   a) The distinguishing electron is $2s^2$ indicating beryllium is $\cdot$ Be $\cdot$
the element and it has 2 valence electrons.
b) The distinguishing electron is $3s^2$ indicating magnesium is $\cdot$ Mg $\cdot$
the element and it has 2 valence electrons.
c) The distinguishing electron is $3p^4$ indicating sulfur is the   : S :
element and it has 6 valence electrons in its outer shell: $3s^2 3p^4$.
d) The distinguishing electron is $4p^2$ indicating germanium is
the element and it has 4 valence electrons in its outer shell: $\cdot$ Ge $\cdot$
$4s^2 3d^{10} 4p^2$. (3d electrons are not considered valence electrons.)

NOTATION FOR IONS (SEC. 7.4)

7.21   a) Li^+ b) P^{3-} c) Br^- d) Ba^{2+}

7.23

	Chemical Symbol	Ion Formed	Number of Electrons in Ion	Number of Protons in Ion
	Ca	Ca^{2+}	18	20
a)	Be	Be^{2+}	2	4
b)	I	I^-	54	53
c)	Al	Al^{3+}	10	13
d)	S	S^{2-}	18	16

7.25   a) (1) A neutral species, number of protons = number of electrons
b) (3) A positively charged species, contains fewer electrons than protons
c) (3) A positively charged species, contains fewer electrons than protons
d) (2) A negatively charged species, contains more electrons than protons

7.27   a) X^{2+} ion contains 10 electrons. The total number of electrons in the neutral atom is 12, (10 + 2).
Mg has 12 electrons, and forms an ion with a 2+ charge.
b) X^{2+} ion contains 18 electrons. The total number of electrons in the neutral atom is 20, (18 + 2).
Ca has 20 electrons, and forms an ion with a 2+ charge.
c) X^{3-} ion contains 18 electrons. The total number of electrons in the neutral atom is 15, (18 − 3).
P has 15 electrons, and forms an ion with a 3− charge.
d) X^- ion contains 10 electrons. The total number of electrons in the neutral atom is 9, (10 − 1).
F has 9 electrons, and forms an ion with a 1− charge.

7.29   a) The period 2 ion with two more protons than electrons, Be^{2+}.
b) The period 2 ion with two fewer protons than electrons, O^{2-}.
c) The period 2 ion with three fewer protons than electrons, N^{3-}.
d) The period 2 ion with four fewer protons than electrons, C^{4-}.

7.31   a) (2), anion   b) (1), cation   c) (3), not an ion   d) (2), anion

THE SIGN AND MAGNITUDE OF IONIC CHARGE (SEC. 7.5)

7.33   a) Nitrogen is predicted to form an ion with a negative charge.
b) Beryllium is predicted to form an ion with a positive charge.
c) Magnesium is predicted to form an ion with a positive charge.
d) Chlorine is predicted to form an ion with a negative charge.

7.35   a) Nitrogen is predicted to form an anion.   b) Beryllium is predicted to form a cation.
c) Magnesium is predicted to form a cation.   d) Chlorine is predicted to form an anion.

7.37   a) Nitrogen is predicted to gain 3 electrons.   b) Beryllium is predicted to lose 2 electrons.
c) Magnesium is predicted to lose 2 electrons.   d) Chlorine is predicted to gain 1 electron.

7.39   a) loss   b) loss   c) loss   d) gain

7.41 a) N^{3-}, a group VA ion contains $(5 + 3)$, a total of 8 valence electrons.

b) P^{3-}, a group VA ion contains $(5 + 3)$, a total of 8 valence electrons.

c) Al^{3+}, a group IIIA ion is formed when aluminum loses 3 valence electrons to obtain a noble gas configuration, with 8 electrons in its new valence shell.

d) Li^+, a group IA ion is formed when lithium loses 1 valence electron to obtain a noble gas configuration, with 2 electrons in its new valence shell.

7.43 Group IIA positive ions have a 2+ charge. To determine the chemical formula, find the number of protons by adding the electrons lost to the number of electrons in the ion.

a) $18, (18 + 2) = 20$ electrons total $= 20$ protons (Ca), 2+ ion is Ca^{2+}

b) $2, (2 + 2) = 4$ electrons total $= 4$ protons (Be), 2+ ion is Be^{2+}

c) $10, (10 + 2) = 12$ electrons total $= 12$ protons (Mg), 2+ ion is Mg^{2+}

d) $36, (36 + 2) = 38$ electrons total $= 38$ protons (Sr), 2+ ion is Sr^{2+}

7.45 To find the chemical symbol of the Group VIIA ion with the same number of electrons as the following positive ions:

1. Determine the number of electrons in each positive ion by subtracting the charge on the positive ion.
2. Subtract 1 from the number of electrons to obtain the identity of the Group VIIA element.

a) $K^+ = (19 - 1) = 18, (18 - 1) = 17$ electrons, Cl, ion is Cl^-

b) $Mg^{2+} = (12 - 2) = 10, (10 - 1) = 9$ electrons, F, ion is F^-

c) $Al^{3+} = (13 - 3) = 10, (10 - 1) = 9$ electrons, F, ion is F^-

d) $Ca^{2+} = (20 - 2) = 18, (18 - 1) = 17$ electrons, Cl, ion is Cl^-

7.47 a) 4 elements form positively charged ions, (Li, Al, Ca, and Rb).

b) 3 elements form anions, (N, F, and Br).

c) 1 element forms an ion with a charge magnitude of 2 (Ca).

d) 2 elements form ions that involve the loss of 2 or more electrons (Ca and Al).

7.49 a) Group IA   b) Group VIA   c) Group VA   d) Group IVA

7.51 Neon has 10 electrons.

a) Na^+ b) F^- c) O^{2-} d) Mg^{2+}

7.53 a) yes, isoelectric   b) no, not isoelectric

c) yes, isoelectric   d) no, not isoelectric

LEWIS STRUCTURES FOR IONIC COMPOUNDS (SEC. 7.6)

7.55

7.57   a) Lithium has 1 valence electron (lose 1); nitrogen has five valence electrons (gain 3).

$$
\begin{array}{ccc}
\text{Li} \cdot & \quad \cdot & \text{Li}^+ \quad \cdot\cdot \quad {}^{3-} \\
\text{Li} \cdot & \cdot \text{N} : \rightarrow & \text{Li}^+ \quad : \text{N} : \\
\text{Li} \cdot & \quad \cdot & \text{Li}^+ \quad \cdot\cdot
\end{array}
$$

b) Mg has 2 valence electrons (lose 2); O has 6 valence electrons (gain 2).

$$
\text{Mg} \cdot \quad \cdot \overset{\displaystyle \cdot \quad \cdot}{\underset{\displaystyle \cdot\cdot}{\text{O}}} : \rightarrow \text{Mg}^{2+} \quad \overset{\displaystyle \cdot\cdot}{\underset{\displaystyle \cdot\cdot}{\text{O}}} :{}^{2-}
$$

c) Cl has 7 valence electrons (gain 1); barium has 2 valence electrons (lose 2). Write Ba first.

$$
\begin{array}{ccc}
\cdot \overset{\cdot\cdot}{\underset{\cdot\cdot}{\text{Cl}}} : & & : \overset{\cdot\cdot}{\underset{\cdot\cdot}{\text{Cl}}} :{}^{-} \\
\text{Ba} \cdot & \rightarrow \text{Ba}^{2+} & \\
\cdot \overset{\cdot\cdot}{\underset{\cdot\cdot}{\text{Cl}}} : & & : \overset{\cdot\cdot}{\underset{\cdot\cdot}{\text{Cl}}} :{}^{-}
\end{array}
$$

d) F has 7 valence electrons (gain 1); K has 1 valence electron (lose 1). Write K first.

$$
\text{K} \cdot \quad \cdot \overset{\cdot\cdot}{\underset{\cdot\cdot}{\text{F}}} : \rightarrow \text{K}^+ \quad : \overset{\cdot\cdot}{\underset{\cdot\cdot}{\text{F}}} :{}^{-}
$$

CHEMICAL FORMULAS FOR IONIC COMPOUNDS (SEC. 7.7)

7.59   a) $CaCl_2$ b) BeO c) AlN d) K_2S

7.61

		F^-	O^{2-}	N^{3-}	C^{4-}
	Na^+	NaF	Na_2O	Na_3N	Na_4C
a)	Ca^{2+}	CaF_2	CaO	Ca_3N_2	Ca_2C
b)	Al^{3+}	AlF_3	Al_2O_3	AlN	Al_4C_3
c)	Ag^+	AgF	Ag_2O	Ag_3N	Ag_4C
d)	Zn^{2+}	ZnF_2	ZnO	Zn_3N_2	Zn_2C

7.63   a) KCl, K is in Group IA, which forms ions with a 1+ charge, K^+.
b) CaS, Ca is in Group IIA, which forms ions with a 2+ charge, Ca^{2+}.
c) MgF_2, Mg is in Group IIA, which forms ions with a 2+ charge, Mg^{2+}.
d) Al_2S_3, Al is in Group IIIA, which forms ions with a 3+ charge, Al^{3+}.

7.65   a) K^+ = 18 electrons, Cl^- = 18 electrons, $\therefore$ they are isoelectronic.
b) Ca^{2+} = 18 electrons, S^{2-} = 18 electrons, $\therefore$ they are isoelectronic.
c) Mg^{2+} = 10 electrons, F^- = 10 electrons, $\therefore$ they are isoelectronic.
d) Al^{3+} = 10 electrons, S^{2-} = 18 electrons, $\therefore$ they are not isoelectronic.

7.67   a) X_2Z b) XZ_3 c) X_3Z d) ZX_2

7.69   a) Be_3N_2 b) $NaBr$ c) SrF_2 d) Al_2S_3

POLYATOMIC-ION-CONTAINING IONIC COMPOUNDS (SEC. 7.9)

7.71 a) CN^- is a polyatomic ion. b) Ca^{2+} is a monoatomic ion.
c) NH_3 is not an ion. d) NH_4^+ is a polyatomic ion.

7.73 a) Al^{3+} and CO_3^{2-} form the compound $Al_2(CO_3)_3$.
b) $Al_2(CO_3)_3$ contains 2 Al^{3+} ions and 3 CO_3^{2-} ions, a total of 5 ions.
c) $Al_2(CO_3)_3$ contains 3 polyatomic ions.
d) $Al_2(CO_3)_3$ contains 2 monoatomic ions.

7.75

		OH^-	CN^-	NO_3^-	SO_4^{2-}
	Na^+	NaOH	NaCN	$NaNO_3$	Na_2SO_4
a)	K^+	KOH	KCN	KNO_3	K_2SO_4
b)	Mg^{2+}	$Mg(OH)_2$	$Mg(CN)_2$	$Mg(NO_3)_2$	$MgSO_4$
c)	Al^{3+}	$Al(OH)_3$	$Al(CN)_3$	$Al(NO_3)_3$	$Al_2(SO_4)_3$
d)	NH_4^+	NH_4OH	NH_4CN	NH_4NO_3	$(NH_4)_2SO_4$

LEWIS STRUCTURES FOR COVALENT COMPOUNDS (SECS. 7.10–7.15)

7.77 a) $: I - I :$ b) $: Cl - F :$ c) $H - S - H$ d) $: F - P - F :$
with $: F :$ below P

7.79 a) $H - Br :$ b) $F - O - F$ c) $: Cl - N - Cl :$ with $: Cl :$ below N d) $: I - Si - I :$ with $: I :$ above and $: I :$ below Si

7.81 a) 1 = H, has 1 valence electron, 6 = S, has 6 valence electrons; 2 hydrogen atoms are needed to make an octet around sulfur, H_2S
b) 2 = C, has 4 valence electrons, 8 = Br, has 7 valence electrons; 4 bromine atoms are needed to make an octet around carbon, CBr_4
c) 3 = N, has 5 valence electrons, 7 = Cl, has 7 valence electrons; 3 chlorine atoms are needed to form an octet around nitrogen, NCl_3
d) 4 = O, has 6 valence electrons, 5 = F, has 7 valence electrons; 2 fluorine atoms are needed to form an octet around oxygen, OF_2

7.83 a) 3 bonding pairs, 2 nonbonding pairs b) 6 bonding pairs, 0 nonbonding pairs
c) 4 bonding pairs, 4 nonbonding pairs d) 7 bonding pairs, 2 nonbonding pairs

7.85 a) 1 triple bond b) 4 single, 1 double c) 2 double d) 2 triple, 1 single

7.87   a) yes   b) yes
        c) no, C has to form 4 bonds   d) yes

7.89   A coordinate covalent bond is a bond in which both of the shared pair electrons come from one of the two atoms.

7.91   a) the N—O bond   b) none
        c) the O—Cl bond   d) the two O—Br bonds

7.93   Resonance structures are two or more Lewis structures for a molecule or ion that have the same arrangement of atoms, contain the same number of electrons, and differ only in the location of the electrons.

7.95

$$\left[\ddot{:O} = N - \ddot{O}: \atop |\atop \ddot{:O}: \right]^{-} \leftrightarrow \left[:\ddot{O} - N = \ddot{O}: \atop |\atop \ddot{:O}: \right]^{-} \leftrightarrow \left[:\ddot{O} - N - \ddot{O}: \atop ||\atop \ddot{:O}: \right]^{-}$$

SYSTEMATIC PROCEDURES FOR DRAWING LEWIS STRUCTURES (SEC. 7.16)

7.97   a) $1 + 5 + (3 \times 6) = 24$ b) $5 + (3 \times 7) = 26$
        c) $5 + (3 \times 6) + 1 = 24$ d) $6 + (4 \times 6) + 2 = 32$

7.99   a) 24 dots   b) 26 dots   c) 24 dots   d) 32 dots

7.101 a) 1 bonding and 3 nonbonding electron pairs   b) 3 bonding and 2 nonbonding electron pairs
        c) 1 bonding and 6 nonbonding electron pairs   d) 1 bonding and 3 nonbonding electron pairs

7.103 a) 1. There are 14 valence electrons
            $(6 \times 1) + (2 \times 4)$.
          2. The two C atoms are central atoms.
          3. Attach hydrogens (all 14 electrons used).

$$\begin{array}{ccc} H & & H \\ | & & | \\ H - C & - & C - H \\ | & & | \\ H & & H \end{array}$$

        b) 1. There are 14 valence electrons
            $(4 \times 1) + (2 \times 5)$.
          2. The two N atoms are central.
          3. Attach H atoms on N's (8 electrons used).
          4. Complete octet on each N by putting a nonbonding
            pair on each N (all 14 electrons used).

$$\begin{array}{ccc} \ddots & & \ddots \\ H - N & - & N - H \\ | & & | \\ H & & H \end{array}$$

        c) 1. There are 20 electrons
            $(2 \times 7) + 4 + (2 \times 1)$.
          2. The C is central.
          3. Attach H and F atoms (8 electrons used).
          4. Complete octet on F (all 20 electrons used).

$$\begin{array}{ccc} & H & \\ & | & \\ :\ddot{F} - & C & - \ddot{F}: \\ & | & \\ & H & \end{array}$$

        d) 1. There are 32 electrons
            $(3 \times 1) + (2 \times 4) + (3 \times 7)$.
          2. The C atoms are central.
          3. Attach H and Cl atoms (12 electrons used).
          4. Complete octets on chlorine (all 32 electrons used).

$$\begin{array}{ccc} & \ddots & \\ H & : Cl : & \\ | & | & \ddots \\ H - C & - C & - Cl : \\ | & | & \ddots \\ H & : Cl : & \\ & \ddots & \end{array}$$

7.105 a) 1. There are 8 electrons
$[5 + (4 \times 1) - 1]$.
2. The N atom is central.
3. Attach H atoms (8 electrons used).

$$\left[\begin{array}{c} H \\ | \\ H-N-H \\ | \\ H \end{array} \right]^{+}$$

b) 1. There are 8 electrons
$[2 + (4 \times 1) + 2]$.
2. The Be atom is central.
3. Attach H atoms (all 8 electrons used).

$$\left[\begin{array}{c} H \\ | \\ H-Be-H \\ | \\ H \end{array} \right]^{2-}$$

c) 1. There are 26 electrons
$[7 + (3 \times 6) + 1]$.
2. The Cl atom is central.
3. Attach O atoms (6 electrons used).
4. Complete octet on O atoms (24 electrons used).
5. Complete octet on Cl with a nonbonding pair (all 26 electrons used).

$$\left[\begin{array}{c} \ddot{\:}O-Cl-O\ddot{\:} \\ | \\ \ddot{\:}O\ddot{\:} \end{array} \right]^{-}$$

d) 1. There are 32 electrons
$[7 + (4 \times 6) + 1]$.
2. The I atom is central.
3. Attach O atoms (8 electrons used).
4. Complete octet on O atoms (all 32 electrons used).

$$\left[\begin{array}{c} \ddot{\:}O\ddot{\:} \\ | \\ \ddot{\:}O-I-O\ddot{\:} \\ | \\ \ddot{\:}O\ddot{\:} \end{array} \right]^{-}$$

7.107 a) 1. There are 24 electrons
$[(2 \times 4) + (2 \times 1) + (2 \times 7)]$.
2. Join 2 C atoms (2 electrons used).
3. Attach H and Cl atoms (8 electrons).
4. Complete octet on Cl atoms (12 electrons used).
5. Complete octet on C atoms by double bond (all 24 electrons used).

$$\begin{array}{c} \ddot{\:}Cl-C=C-H \\ | \qquad | \\ \ddot{\:}Cl\ddot{\:} \quad H \end{array}$$

b) 1. There are 16 electrons
$[(2 \times 4) + (3 \times 1) + 5]$.
2. Join 2 C atoms (2 electrons used).
3. Attach 3 H atoms to a C (6 electrons used).
4. Attach the N atom to other C (2 more electrons used).
5. Complete octet on other C atom by triple bond
(4 electrons used to make a triple bond).
6. Complete octet on N atom (16 electrons used).

$$\begin{array}{c} H \\ | \\ H-C-C \equiv N\ddot{\:} \\ | \\ H \end{array}$$

c) 1. There are 16 electrons
$[(3 \times 4) + (4 \times 1)]$.
2. Join 3 C atoms (4 electrons used).
3. Attach the 4 H atoms (8 electrons).
4. Complete octet on C atoms by triple bond
(all 16 electrons used).

$$\begin{array}{c} H \\ | \\ H-C \equiv C-C-H \\ | \\ H \end{array}$$

d) 1. There are 24 electrons
$[(2 \times 5) + (2 \times 7)]$.
2. Join 2 N atoms (2 electrons).
3. Attach 2 F atoms (4 electrons).
4. Complete octet on 2 F atoms (12 electrons).
5. Complete octet on 2 N atoms by making a double bond
(2 electrons) and 1 nonbonding pair on each (all 24 electrons used).

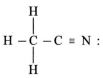

7.109   Follow steps used in Problem 7.97.

a)
$$
\begin{array}{c}
\text{H} \\
| \\
\text{H} - \text{C} - \text{N} = \text{O} \\
| \qquad | \\
\text{H} \quad :\text{O}:
\end{array}
\quad \leftrightarrow \quad
\begin{array}{c}
\text{H} \\
| \\
\text{H} - \text{C} - \text{N} - \text{O}: \\
| \qquad \| \\
\text{H} \quad :\text{O}:
\end{array}
$$

b) $:\text{N} \equiv \text{N} - \ddot{\text{O}}:$ $\leftrightarrow$ $:\ddot{\text{N}} = \text{N} = \ddot{\text{O}}:$ $\leftrightarrow$ $:\ddot{\text{N}} - \text{N} \equiv \text{O}:$

c)
$$
\left[
\begin{array}{c}
\ddot{\text{O}} = \text{C} - \ddot{\text{O}}: \\
| \\
:\ddot{\text{O}}:
\end{array}
\right]^{2-}
\leftrightarrow
\left[
\begin{array}{c}
:\ddot{\text{O}} - \text{C} - \ddot{\text{O}}: \\
\| \\
:\text{O}:
\end{array}
\right]^{2-}
\leftrightarrow
\left[
\begin{array}{c}
:\ddot{\text{O}} - \text{C} = \ddot{\text{O}} \\
| \\
:\ddot{\text{O}}:
\end{array}
\right]^{2-}
$$

d)
$$
\left[
\begin{array}{c}
:\ddot{\text{S}} - \text{C} \equiv \text{N}:
\end{array}
\right]^{-}
\leftrightarrow
\left[
\begin{array}{c}
\ddot{\text{S}} = \text{C} = \ddot{\text{N}}:
\end{array}
\right]^{-}
\leftrightarrow
\left[
\begin{array}{c}
:\text{S} \equiv \text{C} - \ddot{\text{N}}:
\end{array}
\right]^{-}
$$

7.111   Total electrons in structure $= 32$; total valence electrons $= 4(6) + 7 = 31$; $31 - 32 = -1$; $\therefore n = 1$.

MOLECULAR GEOMETRY (VSEPR THEORY) (SEC. 7.17)

7.113
a) The electron arrangement about the central atom involves 2 single bonds and 2 nonbonding locations. This arrangement produces an angular geometry.

b) The electron arrangement about the central atom involves 1 single bond and 1 triple bond, and 0 nonbonding locations. This arrangement produces a linear geometry.

c) The electron arrangement about the central atom involves 1 single bond, 1 double bond, and 1 nonbonding location. This arrangement produces angular geometry.

d) The electron arrangement of one double bond between two atoms, and two nonbonding locations, produces a linear arrangement between two atoms.

7.115
a) The electron arrangement about the central atom involves 2 bonding locations and 2 nonbonding locations. This arrangement produces an angular geometry.

b) The electron arrangement about the central atom involves 2 bonding locations and no nonbonding locations. This arrangement produces a linear geometry.

c) The electron arrangement about the central atom involves 2 bonding locations and no nonbonding locations. This arrangement produces a linear geometry.

d) The electron arrangement about the central atom involves 2 bonding locations and 1 nonbonding location. This arrangement produces an angular geometry.

7.117
a) The electron arrangement about the central atom involves 3 bonding locations and 1 nonbonding location. This arrangement produces a trigonal pyramidal geometry.

b) The electron arrangement about the central atom involves 2 bonding locations and no nonbonding locations. This arrangement produces a linear geometry.

c) The electron arrangement about the central atom involves 2 bonding locations and 1 nonbonding location. This arrangement produces an angular geometry.

d) The electron arrangement about the central atom involves 4 bonding locations and no nonbonding locations. This arrangement produces a tetrahedral geometry.

7.119 a) The electron arrangement about the central atom involves 4 bonding locations and no nonbonding locations. This arrangement produces a tetrahedral geometry.

b) The electron arrangement about the central atom involves 2 bonding locations and no nonbonding locations. This arrangement produces a linear geometry.

c) The electron arrangement about the central atom involves 3 bonding locations and 1 nonbonding location. This arrangement produces a trigonal pyramidal geometry.

d) The electron arrangement about the central atom involves 3 bonding locations and 1 nonbonding location. This arrangement produces a trigonal pyramidal geometry.

7.121 a) Each central atom has 3 bonding locations and 1 nonbonding location, giving a trigonal pyramid for each center. The centers are joined by one axis of each trigonal pyramid.

b) The C has 4 bonding locations and no nonbonding locations and is thus tetrahedral. The O has two bonding locations and 2 nonbonding locations and is thus angular.

ELECTRONEGATIVITY (SEC. 7.18)

7.123 Electronegativity increases across a period and decreases down a group.

7.125 a) Be and O are in the same period; O is farther to the right and therefore more electronegative.

b) Be and Ca are in the same group; Be is above Ca and is therefore more electronegative.

c) H is in the first period and C is in the second period. Since C is much farther to the right, it is more electronegative.

d) Ca is in the second group in the fourth period, while Cs is in the sixth period and in the first group. The larger gap between periods prevails and therefore Ca is more electronegative.

7.127 For representative elements, electronegativity values generally increase from left to right within a period, and from bottom to top within a group. F with a value of 4.0 is the most electronegative element and is used as a reference.

a) Electronegativity increases from bottom to top of a group; N has the highest electronegativity.

b) Electronegativity increases from left to right within a period; C has the highest electronegativity.

c) Electronegativity increases from left to right within a period and bottom to top of a group; B has the highest electronegativity.

d) F is the most electronegative element.

7.129 a) C = 2.5, $\therefore$ elements N (3.0), O (3.5), F (4.0), Cl (3.0), and Br (2.8) are more electronegative.

b) 1.0, $\therefore$ elements that are less electronegative include Na (0.9), K (0.8), Rb (0.8), Cs (0.7), Ba (0.9), Fr (0.7), and Ra (0.9).

c) The 4 most electronegative elements are F (4.0), O (3.5), N and Cl (3.0).

d) For period 2, elements increase from left to right by 0.5 in electronegative values.

7.131 a) True, Be has a higher electronegativity value than Mg.

b) False, N has a higher electronegativity value than B.

c) False, Br has a higher electronegativity value than K.

d) False, F has a higher electronegativity value than As.

BOND POLARITY (SEC. 7.19)

7.133 a) N—S   b) N—Br   c) N—O   d) N—F
    $\delta-$ $\delta+$ $\delta-$ $\delta+$ $\delta+$ $\delta-$ $\delta+$ $\delta-$

7.135 Using the differences in electronegativity between bonding atoms:
a) P—O with a difference of 1.4, has the greatest polarity.
b) C—S with a difference of 0, has the least ionic character.
c) P—O with a difference of 1.4, has the least covalent character.
d) P—O with a difference of 1.4, has the greatest partial positive charge on the less electronegative atom.

7.137 a) The electronegativity difference between hydrogen (2.1) and oxygen (3.5) is 1.4, producing a polar covalent bond.
b) The electronegativity difference between sulfur (2.5) and oxygen (3.5) is 1.0, producing a polar covalent bond.
c) The electronegativity difference between lithium (1.0) and oxygen (3.5) is 2.5, producing an ionic bond.
d) The electronegativity difference between fluorine (4.0) and oxygen (3.5) is 0.5, producing a polar covalent bond.

7.139

	Bond Polarity	More Electronegative Atom	Atom With Negative Partial Charge ($\delta-$)	Bond Type
	N—S	N	N	polar covalent
a)	Se—S	S	S	nonpolar covalent
b)	O—S	O	O	polar covalent
c)	Br—S	Br	Br	nonpolar covalent
d)	F—S	F	F	polar covalent

7.141 a) 0.3 difference: bonds with an electronegativity difference less than 0.4 are nonpolar covalent bonds.
b) 1.2 difference; bonds with an electronegativity difference more than 0.4 but less than 1.5 are considered polar covalent bonds.
c) 1.9 difference; bonds with an electronegativity difference between 1.5 and 2.0 are considered to be ionic if the bond involves a metal and a nonmetal, and polar covalent if the bond involves two nonmetals.
d) 2.1 difference; bonds with an electronegativity difference of 2.0 or greater are considered ionic bonds.

7.143 a) 0.3 difference; bonds have more covalent character.
b) 1.2 difference; bonds have more covalent character.
c) 1.9 difference; bonds have more ionic character.
d) 2.1 difference; bonds have more ionic character.

MOLECULAR POLARITY (SEC. 7.20)

7.145 a) nonpolar   b) polar   c) polar   d) nonpolar

7.147 a) nonpolar   b) polar   c) polar   d) nonpolar

7.149 a) F_2 is nonpolar and BrF is polar.   b) HOCl and HCN are both polar.
c) CH_4 and CCl_4 are both nonpolar.   d) SO_3 is nonpolar and NF_3 is polar.

7.151 a) CH_4 has 4 nonpolar covalent bonds.   b) H_2O has 2 polar covalent bonds.
c) CO_2 has 2 polar covalent bonds.   d) O_3 has 2 nonpolar covalent bonds.

Multi-Concept Problems

7.153 a) Mg: $1s^2 2s^2 2p^6 3s^2$ Mg^{2+}: $1s^2 2s^2 2p^6$
      b) F: $1s^2 2s^2 2p^5$ F^-: $1s^2 2s^2 2p^6$
      c) N: $1s^2 2s^2 2p^3$ N^{3-}: $1s^2 2s^2 2p^6$
      d) Ca^{2+}: $1s^2 2s^2 2p^6 3s^2 3p^6$ S^{2-}: $1s^2 2s^2 2p^6 3s^2 3p^6$

7.155 a) nonisoelectronic cations   b) nonisoelectronic anions
      c) isoelectronic cations   d) nonisoelectronic anions

7.157 a) ionic, Al is a metal, O is a nonmetal   b) molecular, H and O are both nonmetals
      c) ionic, K is a metal, S is a nonmetal   d) molecular, N and H are both nonmetals

7.159 a) and d) are molecules; b) and c) are ionic compounds for which the formula unit is the simplest ratio of ions.

7.161 a) CaO, Ca has charge of $+2$ b) MgF_2, Mg has a charge of $+2$
      c) Na_2S, neither ion contains charge of $+2$ d) NO, neither ion contains charge of $+2$

7.163 a) O   b) N   c) O   d) F

7.165 N—O bonds in N_2O_3 have an electronegativity difference of 0.5, and are considered polar covalent.

7.167 a) correct number of electron dots, but improperly placed
      b) not enough electron dots
      c) correct number of electron dots, but improperly placed
      d) correct number of electron dots, but improperly placed

7.169 Elements: A = 3.8   B = 3.3   C = 2.8   D = 1.3
      Compounds: BA (0.5)   DA (2.5)   DB (2.0)   CA (1.0)
      a) increasing ionic bond character: BA, CA, DB, DA
      b) decreasing covalent bond character: BA, CA, DB, DA

7.171 a) same, both single   b) different, single and triple
      c) different, triple, and double   d) different, triple and double

7.173 a) There are 26 electrons shown. The 3 oxygen and 1 hydrogen provide 19, so X must provide 7: X = Cl
      b) There are 20 electrons shown. The 2 oxygen provide 12, the ionic charge provides 1, so X must provide 7: X = Cl

7.175 a) Ca^{2+} $\left[\begin{array}{c} \ddot{\overset{..}{O}} \\ | \\ \ddot{O} - \overset{..}{S} - \ddot{O} \\ | \\ \ddot{\underset{..}{O}} \end{array} \right]^{2-}$ b) $\left[\begin{array}{c} H \\ | \\ H - N - H \\ | \\ H \end{array} \right]^{+}$ $\left[\begin{array}{c} \ddot{O} - N - \ddot{O} \\ \| \\ \overset{..}{O} \end{array} \right]^{-}$

Note that other resonance structures are possible for the $SO_4{}^{2-}$ and $NO_3{}^-$ ions!

7.177 a) The S in SO_4^{2-} has 4 electron groups around it with a tetrahedral geometry.
  b) The S in HS^- has 4 electron groups around it with a linear geometry.
  c) The S in H_2S has 4 electron groups around it with an angular geometry.
  d) The S in SO_2 has 3 electron groups around it with an angular geometry.

7.179 a) 109.5° because the electron pair arrangement about the oxygen atom is tetrahedral.
  b) 120° because the electron pair arrangement about the carbon atom is trigonal planar.

Answers to Multiple-Choice Practice Test

MC 7.1	a	**MC 7.2**	d	**MC 7.3**	c	**MC 7.4**	b	**MC 7.5**	c	**MC 7.6**	c
MC 7.7	b	**MC 7.8**	e	**MC 7.9**	c	**MC 7.10**	c	**MC 7.11**	c	**MC 7.12**	c
MC 7.13	d	**MC 7.14**	d	**MC 7.15**	d	**MC 7.16**	c	**MC 7.17**	e	**MC 7.18**	b
MC 7.19	d	**MC 7.20**	c								

CHAPTER EIGHT
Chemical Nomenclature

NOMENCLATURE CLASSIFICATIONS FOR COMPOUNDS (SEC. 8.1)

8.1   a) molecular   b) ionic   c) ionic   d) molecular

8.3   a) ionic   b) ionic   c) ionic   d) ionic

8.5   a) yes   b) yes   c) no, both molecular   d) no, both ionic

TYPES OF BINARY IONIC COMPOUNDS (SEC. 8.2)

8.7   a) binary   b) not binary   c) binary   d) binary

8.9   a) not ionic   b) not ionic   c) ionic   d) not ionic

8.11   a) not binary ionic   b) not binary ionic   c) binary ionic   d) not binary ionic

8.13   a) fixed-charge   b) variable-charge   c) fixed-charge   d) variable-charge

8.15   a) and d)

8.17   a) variable-charge   b) variable-charge   c) fixed-charge   d) fixed-charge

NOMENCLATURE FOR BINARY IONIC COMPOUNDS (SEC. 8.3)

8.19   a) Zn^{2+} ion, is from a fixed charge metal, no Roman numeral required.
     b) Co^{3+} ion, is from a variable charge metal, a Roman numeral is required.
     c) Pb^{4+} ion, is from a variable charge metal, a Roman numeral is required.
     d) Ag^{+} ion, is from a fixed charge metal, no Roman numeral required.

8.21   a) Be^{2+}, beryllium ion   b) Pb^{2+}, lead(II) ion
     c) Au^{+}, gold(I) ion   d) Li^{+}, lithium ion

8.23   Roman numerals are required only for variable charge metals.
     a) Br^{-} ion, a nonmetal, it does not require a Roman numeral.
     b) P^{3-} ion, a nonmetal, it does not require a Roman numeral.
     c) S^{2-} ion, a nonmetal, it does not require a Roman numeral.
     d) I^{-} ion, a nonmetal, it does not require a Roman numeral.

8.25   a) Br^{-}, bromide ion   b) P^{3-}, phosphide ion   c) S^{2-}, sulfide ion   d) I^{-}, iodide ion

8.27   a) Ca^{2+}, calcium ion   b) H^{+}, hydrogen ion   c) Fe^{3+}, iron(III) ion   d) As^{3-}, arsenide ion

8.29   a) Zn^{2+} b) Pb^{2+} c) Ca^{2+} d) N^{3-}

8.31   a) fixed-charge   b) variable-charge   c) fixed-charge   d) variable-charge

8.33   a) no Roman numeral needed   b) yes, Roman numeral needed
           c) no Roman numeral needed   d) yes, Roman numeral needed

8.35   a) $AlCl_3$, aluminum chloride   b) $NiCl_3$, nickel(III) chloride
           c) ZnO, zinc oxide   d) CoO, cobalt(II) oxide

8.37   a) Al_2O_3, aluminum oxide   b) CoF_3, cobalt(III) fluoride
           c) Ag_3N, silver nitride   d) BaS, barium sulfide

8.39   a) iron(II) bromide, iron(III) bromide   b) copper(I) sulfide, copper(II) sulfide
           c) tin(II) sulfide, tin(IV) sulfide   d) nickel(II) oxide, nickel(III) oxide

8.41   a) sodium oxide, Na_2S b) copper(II) sulfide, CuS
           c) $AlCl_3$, aluminum chloride   d) $AuCl_3$, gold(III) chloride

8.43   a) plumbic oxide   b) auric chloride   c) iron(III) iodide   d) tin(II) bromide

8.45   In b) and c), both names denote the same compound.

8.47   a) Li_2S b) ZnS   c) Al_2S_3 d) Ag_2S

8.49   a) CuS   b) Cu_3N_2 c) SnO   d) SnO_2

8.51

Formula of Positive Ion	Formula of Negative Ion	Compound Formula	Compound Name
Mg^{2+}	Cl^-	$MgCl_2$	magnesium chloride
Al^{3+}	O^{2-}	Al_2O_3	aluminum oxide
Pb^{2+}	Br^-	$PbBr_2$	lead(II) bromide
Fe^{2+}	S^{2-}	FeS	iron(II) sulfide
Zn^{2+}	Br^-	$ZnBr_2$	zinc bromide

CHEMICAL FORMULAS FOR POLYATOMIC IONS (SEC. 8.4)

8.53   a) OH^- b) NH_4^+ c) NO_3^- d) ClO_4^-

8.55   a) peroxide   b) thiosulfate   c) oxalate   d) chlorate

8.57   a) SO_4^{2-}, SO_3^{2-} b) PO_4^{3-}, HPO_4^{2-} c) OH^-, O_2^{2-} d) CrO_4^{2-}, $Cr_2O_7^{2-}$

NOMENCLATURE FOR POLYATOMIC-ION-CONTAINING COMPOUNDS (SEC. 8.5)

8.59   a) Ag has a charge of +1.   b) Zn has a charge of +2.
           c) Pb has a charge of +2.   d) Cu has a charge of +1.

8.61   a) fixed-charge ionic compound   b) fixed-charge ionic compound
           c) variable-charge ionic compound   d) variable-charge ionic compound

8.63   a) silver nitrate   b) zinc sulfate
           c) lead(II) cyanide   d) copper(I) phosphate

8.65 a) iron(III) carbonate, iron(II) carbonate   b) gold(I) sulfate, gold(III) sulfate
      c) tin(II) hydroxide, tin(IV) hydroxide   d) chromium(III) acetate, chromium(II) acetate

8.67 a) Ag_2CO_3 b) $AuNO_3$ c) $Fe_2(SO_4)_3$ d) CuCN

8.69 a) sodium sulfate, Na_2SO_4 b) copper(II) nitrate, $Cu(NO_3)_2$
      c) $CaCO_3$, calcium carbonate   d) $AuPO_4$, gold(III) phosphate

8.71

Formula of Positive Ion	Formula of Negative Ion	Compound Formula	Compound Name
Mg^{2+}	NO_3^-	$Mg(NO_3)_2$	magnesium nitrate
Al^{3+}	SO_4^{2-}	$Al_2(SO_4)_3$	aluminum sulfate
Cu^{2+}	CN^-	$Cu(CN)_2$	copper(II) cyanide
Fe^{2+}	OH^-	$Fe(OH)_2$	iron(II) hydroxide
Zn^{2+}	N_3^-	$Zn(N_3)_2$	zinc azide

NOMENCLATURE FOR BINARY MOLECULAR COMPOUNDS (SEC. 8.6)

8.73 a) CO: O is to the right in the order given in the textbook on page 306.
      b) Cl_2O: O is to the right in the order given on page 306.
      c) N_2O: O is to the right in the order given on page 306.
      d) HI: I is to the right in the order given on page 306.

8.75 a) 7   b) 5   c) 3   d) 10

8.77 a) tetraphosphorus decoxide   b) sulfur tetrafluoride
      c) carbon tetrabromide   d) nitrogen dioxide

8.79 a) SO   b) S_2O c) SO_2 d) S_7O_2

8.81 a) Carbon dioxide has the chemical formula CO_2.
      b) N_2O_5 has the name dinitrogen pentoxide.
      c) SO_3 has the name sulfur trioxide.
      d) SF_2 has the name sulfur difluoride.

8.83 a) hydrogen sulfide   b) hydrogen fluoride
      c) ammonia   d) methane

8.85 a) N_2O b) HBr   c) C_2H_6 d) H_2Te

NOMENCLATURE FOR ACIDS (SEC. 8.7)

8.87 a) no   b) yes   c) yes   d) no

8.89 a) NO_3^- b) I^- c) ClO^- d) PO_3^{3-}

8.91 a) HCN   b) H_2SO_4 c) HNO_2 d) H_3BO_3

8.93   a) nitric acid   b) hydroiodic acid
         c) hypochlorous acid   d) phosphorous acid

8.95   a) hydrocyanic acid   b) sulfuric acid
         c) nitrous acid   d) boric acid

8.97   a) HCl b) H_2CO_3 c) $HClO_3$ d) H_2SO_4

8.99

Positive Ion Present in Solution	Negative Ion Present in Solution	Chemical Formula of Acid	Name of Acid
H^+	SO_3^{2-}	H_2SO_3	sulfurous acid
H^+	NO_2^-	HNO_2	nitrous acid
H^+	PO_4^{3-}	H_3PO_4	phosphoric acid
H^+	CN^-	HCN	hydrocyanic acid
H^+	ClO^-	$HClO$	hypochlorous acid

8.101 a) arsenous acid   b) periodic acid   c) hypophosphorous acid   d) bromous acid

8.103 a) hydrogen bromide b) hydrocyanic acid   c) hydrogen sulfide   d) hydroiodic acid

8.105 a) Nitrous acid has the chemical formula HNO_2.
         b) Sulfurous acid has the chemical formula H_2SO_3.
         c) HCN has the name hydrocyanic acid.
         d) H_3PO_4 has the name phosphoric acid.

SYSTEMATIC PROCEDURES FOR USING NOMENCLATURE RULES (SEC. 8.8)

8.107 a) HNO_3, (4) molecular acid
         b) Na_2SO_4, (2) polyatomic-ion-containing ionic
         c) SO_3, (3) molecular nonacid
         d) $BaCl_2$, (1) binary ionic

8.109 a) HNO_3, nitric acid   b) Na_2SO_4, sodium sulfate
         c) SO_3, sulfur trioxide   d) $BaCl_2$, barium chloride

8.111 a) CaI_2, calcium iodide   b) $HCl(aq)$, hydrochloric acid
         c) Cu_2SO_4, copper(I) sulfate d) N_2O, dinitrogen monoxide or nitrous oxide

8.113 a) Ca_3N_2, calcium nitride   b) $Ca(NO_3)_2$, calcium nitrate
         c) $Ca(NO_2)_2$, calcium nitrite   d) $Ca(CN)_2$, calcium cyanide

8.115 a) HF b) MgS c) Li_3N d) SBr_2

8.117 a) no   b) no   c) yes   d) no

8.119 a) yes   b) no   c) no   d) yes

8.121 a) iron(II) hydroxide, $Fe(OH)_2$ b) silver bromide, $AgBr$
         c) nitrogen trifluoride, NF_3 d) chlorous acid, $HClO_2$

8.123 a) K_3P b) K_3PO_4 c) K_2HPO_4 d) KH_2PO_4

8.125 a) Cl_2O, CO_2 b) NO_2, SO_2 c) SF_2, SCl_2 d) N_2O_3

8.127 a) $CaCO_3$, HNO_2 b) $NaClO_4$, $NaClO_3$ c) $HClO_2$, $HClO$ d) Li_2CO_3, Li_3PO_4

Multi-Concept Problems

8.129 a) Cu_2SO_4 has the name copper(I) sulfate. b) $MgCl_2$ has the name magnesium chloride.
c) CO has the name carbon monoxide. d) N_2O_5 has the name dinitrogen pentoxide.

8.131 a) Silver phosphate, Ag_3PO_4 contains 3 Ag^+ ions and 1 PO_4^{3-} ion, a total of 4 ions.
b) Zinc oxide, ZnO contains 1 Zn^{2+} ion and 1 O^{2-} ion, a total of 2 ions.
c) Dinitrogen monoxide, N_2O is a binary molecular compound and does not contain any ions.
d) Ammonium nitrate, NH_4NO_3 contains 1 NH_4^+ ion and 1 NO_3^- ion, a total of 2 ions.

8.133 • Element X forms the ionic compound CaX, meaning X forms an ion with a 2^- charge and is in group VIA.
• Element Q forms the molecular compound NQ_3, meaning each Q shares one valence electron with nitrogen and is in group VIIA.
• If both X and Y are period 4 elements? X must be selenium and Q must be bromine, forming compound $SeBr_2$, selenium dibromide.

8.135 a) sodium nitrate
b) aluminum sulfide, magnesium nitride, beryllium phosphide
c) iron(III) oxide
d) gold(I) chlorate

8.137 a) $Ni_2(SO_4)_3$ b) Ni_2O_3 c) $Ni_2(C_2O_4)_3$ d) $Ni(NO_3)_3$

8.139 The superoxide ion is O_2^-. The nitronium ion is NO_2^+. The formula for nitronium superoxide is NO_2O_2.

Answers to Multiple-Choice Practice Test

MC 8.1   d   MC 8.2   a   MC 8.3   d   MC 8.4   c   MC 8.5   b   MC 8.6   c

MC 8.7   d   MC 8.8   c   MC 8.9   a   MC 8.10 a   MC 8.11 a   MC 8.12 b

MC 8.13 c   MC 8.14 e   MC 8.15 d   MC 8.16 e   MC 8.17 a   MC 8.18 a

MC 8.19 c   MC 8.20 d

CHAPTER NINE
Chemical Calculations: The Mole Concept and Chemical Formulas

LAW OF DEFINITE PROPORTIONS (SEC. 9.1)

9.1   The same: 42.9% C and 57.1% O. Any size sample of CO will have that composition.

9.3   According to the law of definite proportions, the percentage A in each sample should be the same and the percentage D should also be the same.

Sample I: $\%A = \dfrac{10.03 \text{ g}}{17.35 \text{ g}} \times 100 = 57.809798 \text{ (calc)} = 57.81\% \text{ (corr)}$

                   $\%D = \dfrac{7.32 \text{ g}}{17.35 \text{ g}} \times 100 = 42.190202 \text{ (calc)} = 42.2\% \text{ (corr)}$

Sample II: $\%A = \dfrac{13.17 \text{ g}}{22.78 \text{ g}} \times 100 = 57.813872 \text{ (calc)} = 57.81\% \text{ (corr)}$

                   $\%D = \dfrac{9.61 \text{ g}}{22.78 \text{ g}} \times 100 = 42.186128 \text{ (calc)} = 42.2\% \text{ (corr)}$

9.5   The mass ratio between X and Q present in the sample will be the same for samples of the same compound.

Experiment	Q/X mass ratio
1	$\dfrac{8.90 \text{ g}}{3.37 \text{ g}} = 2.6409496 \text{ (calc)} = 2.64 \text{ (corr)}$
2	$\dfrac{1.711 \text{ g}}{0.561 \text{ g}} = 3.0499109 \text{ (calc)} = 3.05 \text{ (corr)}$
3	$\dfrac{71.0 \text{ g}}{26.9 \text{ g}} = 2.6394052 \text{ (calc)} = 2.64 \text{ (corr)}$

∴ Experiments 1 and 3 produced the same compound.

9.7   If all 50.1 g S is used, the mass of SO_2 produced would be

$50.1 \text{ g S} \times \dfrac{100 \text{ g } SO_2}{50.1 \text{ g S}} = 100 \text{ (calc)} = 100. \text{ g } SO_2 \text{ (corr)}$

If all 75.0 g O is used, the mass of SO_2 produced would be

$75.0 \text{ g O} \times \dfrac{100 \text{ g } SO_2}{49.1 \text{ g O}} = 152.7494908 \text{ (calc)} = 153 \text{ g } SO_2 \text{ (corr)}$

But after making 100. g SO_2, all the S is gone. ∴ The maximum possible is 100. g SO_2.

FORMULA MASSES (SECS. 9.2–9.3)

9.9 a)
	H	$1 \times$ 1.01 amu $=$	1.01 amu
	Cl	$1 \times$ 35.45 amu $=$	35.45 amu
	O	$1 \times$ 16.00 amu $=$	16.00 amu
	$HClO$ $=$		52.46 amu

b)
	H	$1 \times$ 1.01 amu $=$	1.01 amu
	Cl	$1 \times$ 35.45 amu $=$	35.45 amu
	O	$3 \times$ 16.00 amu $=$	48.00 amu
	$HClO_3$ $=$		84.46 amu

c)
	H	$2 \times$ 1.01 amu $=$	2.02 amu
	S	$1 \times$ 32.06 amu $=$	32.06 amu
	O	$4 \times$ 16.00 amu $=$	64.00 amu
	H_2SO_4 $=$		98.08 amu

d)
	H	$2 \times$ 1.01 amu $=$	2.02 amu
	S	$1 \times$ 32.06 amu $=$	32.06 amu
	O	$3 \times$ 16.00 amu $=$	48.00 amu
	H_2SO_3 $=$		82.08 amu

9.11 a)
	C	$1 \times$ 12.01 amu $=$	12.01 amu
	O	$2 \times$ 16.00 amu $=$	32.00 amu
	CO_2 $=$		44.01 amu

b)
	N	$2 \times$ 14.01 amu $=$	28.02 amu
	O	$1 \times$ 16.00 amu $=$	16.00 amu
	N_2O $=$		44.02 amu

c)
	N	$2 \times$ 14.01 amu $=$	28.02 amu
	H	$4 \times$ 1.01 amu $=$	4.04 amu
	C	$1 \times$ 12.01 amu $=$	12.01 amu
	NH_4CN $=$		44.07 amu

d)
	C	$3 \times$ 12.01 amu $=$	36.03 amu
	H	$8 \times$ 1.01 amu $=$	8.08 amu
	C_3H_8 $=$		44.11 amu

9.13 a)
	Pt	$1 \times$ 195.08 amu $=$	195.08 amu
	Br	$4 \times$ 79.90 amu $=$	319.60 amu
	$PtBr_4$ $=$		514.68 amu

b)
	Au	$1 \times$ 196.97 amu $=$	196.97 amu
	I	$3 \times$ 126.90 amu $=$	380.07 amu
	AuI_3 $=$		577.67 amu

c)
	Os	$3 \times$ 190.23 amu $=$	570.69 amu
	C	$12 \times$ 12.01 amu $=$	144.12 amu
	O	$12 \times$ 16.00 amu $=$	192.00 amu
	$Os_3(CO)_{12}$ $=$		906.81 amu

d) U 4×238.03 amu $=$ 952.12 amu
    O $9 \times$ 16.00 amu $=$ <u>144.00 amu</u>
                  $U_4O_9 =$ 1096.12 amu

9.15 a) Mg 3×24.31 amu $=$ 72.93 amu
        Si 4×28.09 amu $=$ 112.36 amu
        O 12×16.00 amu $=$ 192.00 amu
        H $2 \times$ 1.01 amu $=$ <u>2.02 amu</u>
        $Mg_3(Si_2O_5)_2(OH)_2 =$ 379.31 amu

    b) Ca 3×40.08 amu $=$ 120.24 amu
        Be $1 \times$ 9.01 amu $=$ 9.01 amu
        O 12×16.00 amu $=$ 192.00 amu
        H $2 \times$ 1.01 amu $=$ 2.02 amu
        Si 3×28.09 amu $=$ <u>84.27 amu</u>
        $Ca_3(BeOH)_2(Si_3O_{10}) =$ 407.54 amu

9.17 a) ethylene glycol $= C_2H_6O_2$
        C 2×12.01 amu $=$ 24.02 amu
        H $6 \times$ 1.01 amu $=$ 6.06 amu
        O 2×16.00 amu $=$ <u>32.00 amu</u>
                $C_2H_6O_2 =$ 62.08 amu

    b) lactic acid $= C_3H_6O_3$
        C 3×12.01 amu $=$ 36.03 amu
        H $6 \times$ 1.01 amu $=$ 6.06 amu
        O 3×16.00 amu $=$ <u>48.00 amu</u>
              $C_3H_6O_3 =$ 90.09 amu

9.19 $y(12.01 \text{ amu}) + 8(1.01 \text{ amu}) + 32.06 \text{ amu} = 88.18 \text{ amu}$
              $(12.01) \, y \text{ amu} + 40.14 \text{ amu} = 88.18 \text{ amu}$
                      $(12.01) \, y \text{ amu} = 48.04 \text{ amu}$
                              $y = 4$

9.21 $\dfrac{2 \text{ atoms N}}{1 \text{ molecule}} \times \dfrac{14.01 \text{ amu N}}{1 \text{ atom N}} \times \dfrac{100 \text{ amu molecule}}{63.6 \text{ amu N}} = 44.0566038$ (calc)

                                              $= 44.1$ amu molecule (corr)

9.23 Percent O $= 100 - 63.6 = 36.4\%$ O

    44.1 amu molecule $\times \dfrac{36.4 \text{ amu O}}{100 \text{ amu molecule}} \times \dfrac{1 \text{ atom O}}{16.00 \text{ amu O}} = 1.003275$ (calc)

                                             $= 1$ atom O (corr) and formula is N_2O

PERCENT COMPOSITION (SEC. 9.4)

9.25 a) HClO $= 1.01$ amu $+ 35.45$ amu $+ 16.00$ amu $= 52.46$ amu total

        $\%H = \dfrac{1.01 \text{ amu H}}{52.46 \text{ amu total}} \times 100 = 1.925276401$ (corr) $= 1.93\%$ H (corr)

        $\%Cl = \dfrac{35.45 \text{ amu Cl}}{52.46 \text{ amu total}} \times 100 = 67.57529546$ (corr) $= 67.58\%$ Cl (corr)

        $\%O = \dfrac{16.00 \text{ amu O}}{52.46 \text{ amu total}} \times 100 = 30.49942814$ (corr) $= 30.50\%$ O (corr)

b) $HClO_2$ = 1.01 amu + 35.45 amu + 2(16.00 amu) = 68.46 amu total

$$\%H = \frac{1.01 \text{ amu H}}{68.46 \text{ amu total}} \times 100 = 1.475314052 \text{ (corr)} = 1.48\% \text{ H (corr)}$$

$$\%Cl = \frac{35.45 \text{ amu Cl}}{68.46 \text{ amu total}} \times 100 = 51.78206252 \text{ (corr)} = 51.78\% \text{ Cl (corr)}$$

$$\%O = \frac{32.00 \text{ amu O}}{68.46 \text{ amu total}} \times 100 = 46.74262343 \text{ (corr)} = 46.74\% \text{ O (corr)}$$

c) $HClO_3$ = 1.01 amu + 35.45 amu + 3(16.00 amu) = 84.46 amu total

$$\%H = \frac{1.01 \text{ amu H}}{84.46 \text{ amu total}} \times 100 = 1.195832347 \text{ (corr)} = 1.20\% \text{ H (corr)}$$

$$\%Cl = \frac{35.45 \text{ amu Cl}}{84.46 \text{ amu total}} \times 100 = 41.97253138 \text{ (corr)} = 41.97\% \text{ Cl (corr)}$$

$$\%O = \frac{48.00 \text{ amu O}}{84.46 \text{ amu total}} \times 100 = 56.83163628 \text{ (corr)} = 56.83\% \text{ O (corr)}$$

d) $HClO_4$ = 1.01 amu + 35.45 amu + 4(16.00 amu) = 100.46 amu total

$$\%H = \frac{1.01 \text{ amu H}}{100.46 \text{ amu total}} \times 100 = 1.005375274 \text{ (corr)} = 1.01\% \text{ H (corr)}$$

$$\%Cl = \frac{35.45 \text{ amu Cl}}{100.46 \text{ amu total}} \times 100 = 35.28767669 \text{ (corr)} = 35.29\% \text{ Cl (corr)}$$

$$\%O = \frac{64.00 \text{ amu O}}{100.46 \text{ amu total}} \times 100 = 63.70694804 \text{ (corr)} = 63.71\% \text{ O (corr)}$$

9.27   a) H_2SO_4 = 2(1.01 amu) + 32.06 amu + 4(16.00 amu) = 98.08 amu total

$$\%H = \frac{2.02 \text{ amu H}}{98.08 \text{ amu total}} \times 100 = 2.05954323 \text{ (corr)} = 2.06\% \text{ H (corr)}$$

$$\%S = \frac{32.06 \text{ amu S}}{98.08 \text{ amu total}} \times 100 = 32.68760196 \text{ (corr)} = 32.69\% \text{ S (corr)}$$

$$\%O = \frac{64.00 \text{ amu O}}{98.08 \text{ amu total}} \times 100 = 65.25285481 \text{ (corr)} = 65.25\% \text{ O (corr)}$$

b) $BeSO_4$ = 9.01 amu + 32.06 amu + 4(16.00 amu) = 105.07 amu total

$$\%Be = \frac{9.01 \text{ amu Be}}{105.07 \text{ amu total}} \times 100 = 8.575235557 \text{ (corr)} = 8.58\% \text{ Be (corr)}$$

$$\%S = \frac{32.06 \text{ amu S}}{105.07 \text{ amu total}} \times 100 = 30.51299134 \text{ (corr)} = 30.51\% \text{ S (corr)}$$

$$\%O = \frac{64.00 \text{ amu O}}{105.07 \text{ amu total}} \times 100 = 60.9117731 \text{ (corr)} = 60.91\% \text{ O (corr)}$$

c) $Al_2(SO_4)_3$ = 2(26.98 amu) + 3(32.06 amu) + 12(16.00 amu) = 342.14 amu total

$$\%Al = \frac{53.96 \text{ amu Al}}{342.14 \text{ amu total}} \times 100 = 15.77132168 \text{ (corr)} = 15.77\% \text{ Al (corr)}$$

$$\%S = \frac{96.18 \text{ amu S}}{342.14 \text{ amu total}} \times 100 = 28.11129947 \text{ (corr)} = 28.11\% \text{ S (corr)}$$

$$\%O = \frac{192.00 \text{ amu O}}{342.14 \text{ amu total}} \times 100 = 56.11737885 \text{ (corr)} = 56.12\% \text{ O (corr)}$$

d) $(NH_4)_2SO_4 = 2(14.01 \text{ amu}) + 8(1.01 \text{ amu}) + 32.06 \text{ amu} + 4(16.00 \text{ amu}) = 132.16 \text{ amu total}$

$$\%N = \frac{28.02 \text{ amu N}}{132.16 \text{ amu total}} \times 100 = 21.20157385 \text{ (corr)} = 21.20\% \text{ N (corr)}$$

$$\%H = \frac{8.08 \text{ amu H}}{132.16 \text{ amu total}} \times 100 = 6.113801453 \text{ (corr)} = 6.11\% \text{ H (corr)}$$

$$\%S = \frac{32.06 \text{ amu S}}{132.16 \text{ amu total}} \times 100 = 24.25847458 \text{ (corr)} = 24.26\% \text{ S (corr)}$$

$$\%O = \frac{64.00 \text{ amu O}}{132.16 \text{ amu total}} \times 100 = 48.42615012 \text{ (corr)} = 48.43\% \text{ O (corr)}$$

9.29 $C_{55}H_{72}MgN_4O_5 = 55(12.01 \text{ amu}) + 72(1.01 \text{ amu}) + 24.31 \text{ amu} + 4(14.01 \text{ amu}) + 5(16.00) \text{ amu}$
$= 893.62 \text{ amu total}$

$$\%C = \frac{660.55 \text{ amu C}}{893.62 \text{ amu total}} \times 100 = 73.918444 \text{ (calc)} = 73.92\% \text{ C (corr)}$$

$$\%H = \frac{72.72 \text{ amu H}}{893.61 \text{ amu total}} \times 100 = 8.137687 \text{ (calc)} = 8.14\% \text{ H (corr)}$$

$$\%Mg = \frac{24.31 \text{ amu Mg}}{893.62 \text{ amu total}} \times 100 = 2.7203957 \text{ (calc)} = 2.720\% \text{ Mg (corr)}$$

$$\%N = \frac{56.04 \text{ amu N}}{893.62 \text{ amu total}} \times 100 = 6.2711220 \text{ (calc)} = 6.271\% \text{ N (corr)}$$

$$\%O = \frac{80.00 \text{ amu O}}{893.62 \text{ amu total}} \times 100 = 8.9523511 \text{ (calc)} = 8.952\% \text{ O (corr)}$$

9.31 $\%C \text{ in } CS_2 = \dfrac{12.01 \text{ g C}}{76.13 \text{ g } CS_2} \times 100 = 15.7756469 \text{ (calc)} = 15.78\% \text{ C (corr)}$

$\%S \text{ in } CS_2 = 100 - 15.78 = 84.22\% \text{ S, therefore the percent C and S match.}$

9.33 For C_2H_2, $\%C = \dfrac{2(12.01) \text{ g C}}{2(12.01) + 2(1.01) \text{ g total}} \times 100 = 92.242704 \text{ (calc)} = 92.24\% \text{ C (corr)}$

$\%H = \dfrac{2(1.01) \text{ g H}}{2(12.01) + 2(1.01) \text{ g total}} \times 100 = 7.757296 \text{ (calc)} = 7.76\% \text{ H (corr)}$

For C_6H_6, $\%C = \dfrac{6(12.01) \text{ g C}}{6(12.01) + 2(1.01) \text{ g total}} \times 100 = 92.242704 \text{ (calc)} = 92.24\% \text{ C (corr)}$

$\%H = \dfrac{6(1.01) \text{ g H}}{6(12.01) + 2(1.01) \text{ g total}} \times 100 = 7.757296 \text{ (calc)} = 7.76\% \text{ H (corr)}$

Both compounds have the same ratio of carbon to hydrogen. Thus, the %C and %H will be the same in each.

9.35 $\%N = 100 - 12.6 = 87.4\% \text{ N}$ $52.34 \text{ g H} \times \dfrac{87.4 \text{ g N}}{12.6 \text{ g H}} = 363.0568254 \text{ (calc)} = 364 \text{ g N (corr)}$

THE MOLE: THE CHEMIST'S COUNTING UNIT (SEC. 9.5)

9.37 a) $1.00 \text{ mole Ag} \times \dfrac{6.022 \times 10^{23} \text{ Ag atoms}}{1 \text{ mole Ag}} = 6.022 \times 10^{23} \text{ (calc)} = 6.02 \times 10^{23} \text{ Ag atoms (corr)}$

b) $1.00 \text{ mole } H_2O \times \dfrac{6.022 \times 10^{23} \text{ } H_2O \text{ molecules}}{1 \text{ mole } H_2O} = 6.022 \times 10^{23} \text{ (calc)}$

$= 6.02 \times 10^{23} \text{ } H_2O \text{ molecules (corr)}$

c) $1.00 \text{ mole NaNO}_3 \times \dfrac{6.022 \times 10^{23} \text{ NaNO}_3 \text{ formula units}}{1 \text{ mole NaNO}_3} = 6.022 \times 10^{23} \text{ (calc)}$

$$= 6.02 \times 10^{23} \text{ NaNO}_3 \text{ formula units (corr)}$$

d) $1.00 \text{ mole SO}_4^{2-} \text{ ions} \times \dfrac{6.022 \times 10^{23} \text{ SO}_4^{2-} \text{ ions}}{1 \text{ mole SO}_4^{2-} \text{ ions}} = 6.022 \times 10^{23} \text{ (calc)}$

$$= 6.02 \times 10^{23} \text{ SO}_4^{2-} \text{ ions (corr)}$$

9.39   a) $3.20 \text{ moles Si} \times \dfrac{6.022 \times 10^{23} \text{ Si atoms}}{1 \text{ mole Si}} = 1.92704 \times 10^{24} \text{ (calc)} = 1.93 \times 10^{24} \text{ Si atoms (corr)}$

b) $0.36 \text{ mole Si} \times \dfrac{6.022 \times 10^{23} \text{ Si atoms}}{1 \text{ mole Si}} = 2.16792 \times 10^{23} \text{ (calc)} = 2.2 \times 10^{23} \text{ Si atoms (corr)}$

c) $1.21 \text{ moles Si} \times \dfrac{6.022 \times 10^{23} \text{ Si atoms}}{1 \text{ mole Si}} = 7.28662 \times 10^{23} \text{ (calc)} = 7.29 \times 10^{23} \text{ Si atoms (corr)}$

d) $16.7 \text{ moles Si} \times \dfrac{6.022 \times 10^{23} \text{ Si atoms}}{1 \text{ mole Si}} = 1.005674 \times 10^{25} \text{ (calc)} = 1.01 \times 10^{25} \text{ Si atoms (corr)}$

9.41   a) $3.20 \text{ moles CO}_2 \times \dfrac{6.022 \times 10^{23} \text{ molecules CO}_2}{1 \text{ mole CO}_2} = 1.92704 \times 10^{24} \text{ (calc)}$

$$= 1.93 \times 10^{24} \text{ molecules CO}_2 \text{ (corr)}$$

b) $0.36 \text{ mole CO}_2 \times \dfrac{6.022 \times 10^{23} \text{ molecules CO}_2}{1 \text{ mole CO}_2} = 2.16792 \times 10^{23} \text{ (calc)}$

$$= 2.2 \times 10^{23} \text{ molecules CO}_2 \text{ (corr)}$$

c) $1.21 \text{ moles CO}_2 \times \dfrac{6.022 \times 10^{23} \text{ molecules CO}_2}{1 \text{ mole CO}_2} = 7.28662 \times 10^{23} \text{ (calc)}$

$$= 7.29 \times 10^{23} \text{ molecules CO}_2 \text{ (corr)}$$

d) $16.7 \text{ moles CO}_2 \times \dfrac{6.022 \times 10^{23} \text{ molecules CO}_2}{1 \text{ mole CO}_2} = 1.005674 \times 10^{25} \text{ (calc)}$

$$= 1.01 \times 10^{25} \text{ molecules CO}_2 \text{ (corr)}$$

9.43   a) $1.50 \text{ moles CO}_2 \times \dfrac{6.022 \times 10^{23} \text{ molecules CO}_2}{1 \text{ mole CO}_2} = 9.033 \times 10^{23} \text{ (calc)}$

$$= 9.03 \times 10^{23} \text{ molecules CO}_2 \text{ (corr)}$$

b) $0.500 \text{ mole NH}_3 \times \dfrac{6.022 \times 10^{23} \text{ molecules NH}_3}{1 \text{ mole NH}_3} = 3.011 \times 10^{23} \text{ (calc)}$

$$= 3.01 \times 10^{23} \text{ molecules NH}_3 \text{ (corr)}$$

c) $2.33 \text{ moles PF}_3 \times \dfrac{6.022 \times 10^{23} \text{ molecules PF}_3}{1 \text{ mole PF}_3} = 1.40313 \times 10^{24} \text{ (calc)}$

$= 1.40 \times 10^{24} \text{ molecules PF}_3 \text{ (corr)}$

d) $1.115 \text{ moles N}_2\text{H}_4 \times \dfrac{6.022 \times 10^{23} \text{ molecules N}_2\text{H}_4}{1 \text{ mole N}_2\text{H}_4} = 6.71453 \times 10^{23} \text{ (calc)}$

$= 6.715 \times 10^{23} \text{ molecules N}_2\text{H}_4 \text{ (corr)}$

MOLAR MASS (SEC. 9.6)

9.45 a) $1.000 \text{ mole Ca} \times \dfrac{40.08 \text{ g Ca}}{1 \text{ mole Ca}} = 40.08 \text{ g Ca (calc and corr)}$

b) $1.000 \text{ mole Si} \times \dfrac{28.09 \text{ g Si}}{1 \text{ mole Si}} = 28.09 \text{ g Si (calc and corr)}$

c) $1.000 \text{ mole Co} \times \dfrac{58.93 \text{ g Co}}{1 \text{ mole Co}} = 58.93 \text{ g Co (calc and corr)}$

d) $1.000 \text{ mole Ag} \times \dfrac{107.87 \text{ g Ag}}{1 \text{ mole Ag}} = 107.87 \text{ (calc)} = 107.9 \text{ g Ag (corr)}$

9.47 Molar masses of each element:
a) Ca, 40.08 g/mole
b) Si, 28.09 g/mole
c) Co, 58.93 g/mole
d) Ag, 107.87 g/mole

9.49 Mass of 1.000 mole each compound:

a) $1.000 \text{ mole CO}_2 \times \dfrac{44.01 \text{ g CO}_2}{1 \text{ mole CO}_2} = 44.01 \text{ g CO}_2 \text{ (calc and corr)}$

b) $1.000 \text{ mole SO}_2 \times \dfrac{64.06 \text{ g SO}_2}{1 \text{ mole SO}_2} = 64.06 \text{ g SO}_2 \text{ (calc and corr)}$

c) $1.000 \text{ mole NO}_2 \times \dfrac{46.01 \text{ g NO}_2}{1 \text{ mole NO}_2} = 46.01 \text{ g NO}_2 \text{ (calc and corr)}$

d) $1.000 \text{ mole N}_2\text{H}_4 \times \dfrac{32.06 \text{ g N}_2\text{H}_4}{1 \text{ mole N}_2\text{H}_4} = 32.06 \text{ g N}_2\text{H}_4 \text{ (calc and corr)}$

9.51 a) $\text{CO}_2 = 12.01 \text{ amu} + 2(16.00 \text{ amu}) = 44.01 \text{ amu (calc and corr)}$
mass of 1 molecule CO_2 $= 44.01 \text{ amu}$
molar mass of CO_2 molecule $= 44.01 \text{ g/mole}$

b) $\text{SO}_2 = 32.06 \text{ amu} + 2(16.00 \text{ amu}) = 64.06 \text{ amu (calc and corr)}$
mass of 1 molecule SO_2 $= 64.06 \text{ amu}$
molar mass of SO_2 molecule $= 64.06 \text{ g/mole}$

c) $\text{NO}_2 = 14.01 \text{ amu} + 2(16.00 \text{ amu}) = 46.01 \text{ amu (calc and corr)}$
mass of 1 molecule NO_2 $= 46.01 \text{ amu}$
molar mass of NO_2 molecule $= 46.01 \text{ g/mole}$

d) $\text{N}_2\text{H}_4 = 2(14.01 \text{ amu}) + 4(1.01 \text{ amu}) = 32.06 \text{ amu (calc and corr)}$
mass of 1 molecule N_2H_4 $= 32.06 \text{ amu}$
molar mass of N_2H_4 molecule $= 32.06 \text{ g/mole}$

9.53　a)　$2.314 \text{ moles Al(OH)}_3 \times \dfrac{78.01 \text{ g Al(OH)}_3}{1 \text{ mole Al(OH)}_3} = 180.51514 \text{ (calc)} = 180.5 \text{ g Al(OH)}_3 \text{ (corr)}$

　　　b)　$2.314 \text{ moles Mg}_3\text{N}_2 \times \dfrac{100.95 \text{ g Mg}_3\text{N}_2}{1 \text{ mole Mg}_3\text{N}_2} = 233.5983 \text{ (calc)} = 233.6 \text{ g Mg}_3\text{N}_2 \text{ (corr)}$

　　　c)　$2.314 \text{ moles Cu(NO}_3)_2 \times \dfrac{187.57 \text{ g Cu(NO}_3)_2}{1 \text{ mole Cu(NO}_3)_2} = 434.03698 \text{ (calc)} = 434.0 \text{ g Cu(NO}_3)_2 \text{ (corr)}$

　　　d)　$2.314 \text{ moles La}_2\text{O}_3 \times \dfrac{325.82 \text{ g La}_2\text{O}_3}{1 \text{ mole La}_2\text{O}_3} = 753.94748 \text{ (calc)} = 753.9 \text{ g La}_2\text{O}_3 \text{ (corr)}$

9.55　a)　$2.00 \text{ moles Cu} \times \dfrac{63.55 \text{ g Cu}}{1 \text{ mole Cu}} = 127.1 \text{ (calc)} = 127 \text{ g Cu (corr)}$

　　　　　$2.00 \text{ moles O} \times \dfrac{16.00 \text{ g O}}{1 \text{ mole O}} = 32.00 \text{ (calc)} = 32.0 \text{ g O (corr)}$　$\therefore 2.00 \text{ mole Cu greater}$

　　　b)　$1.00 \text{ mole Br} \times \dfrac{79.90 \text{ g Br}}{1 \text{ mole Br}} = 79.90 \text{ (calc)} = 79.9 \text{ g Br (corr)}$

　　　　　$5.00 \text{ moles Be} \times \dfrac{9.01 \text{ g Be}}{1 \text{ mole Be}} = 45.05 \text{ (calc)} = 45.0 \text{ g Be (corr)}$　$\therefore 1.00 \text{ mole Br greater}$

　　　c)　$2.00 \text{ moles CO} \times \dfrac{28.01 \text{ g CO}}{1 \text{ mole CO}} = 56.02 \text{ (calc)} = 56.0 \text{ g CO (corr)}$

　　　　　$1.50 \text{ moles N}_2\text{O} \times \dfrac{44.02 \text{ g N}_2\text{O}}{1 \text{ mole N}_2\text{O}} = 66.03 \text{ (calc)} = 66.0 \text{ g N}_2\text{O (corr)}$　$\therefore 1.50 \text{ mole N}_2\text{O greater}$

　　　d)　$4.87 \text{ moles B}_2\text{H}_6 \times \dfrac{27.68 \text{ g B}_2\text{H}_6}{1 \text{ mole B}_2\text{H}_6} = 134.802 \text{ (calc)} = 135 \text{ g B}_2\text{H}_6 \text{ (corr)}$

　　　　　$0.35 \text{ mole U} \times \dfrac{238.03 \text{ g U}}{1 \text{ mole U}} = 83.31 \text{ (calc)} = 83 \text{ g U (corr)}$　$\therefore 4.87 \text{ moles B}_2\text{H}_6 \text{ greater}$

9.57　Molar mass has units of $\dfrac{\text{g}}{\text{mole}}$, so take the number of grams and divide by the number of moles.

　　　a)　$\dfrac{11.23 \text{ g}}{0.6232 \text{ mole}} = 18.0198973 \text{ (calc)} = 18.02 \dfrac{\text{g}}{\text{mole}} \text{ (corr)}$

　　　b)　$\dfrac{20.14 \text{ g}}{0.5111 \text{ mole}} = 39.40520446 \text{ (calc)} = 39.41 \dfrac{\text{g}}{\text{mole}} \text{ (corr)}$

　　　c)　$\dfrac{352.6 \text{ g}}{2.357 \text{ moles}} = 149.5969453 \text{ (calc)} = 149.6 \dfrac{\text{g}}{\text{mole}} \text{ (corr)}$

　　　d)　$\dfrac{100.0 \text{ g}}{1.253 \text{ moles}} = 79.8084597 \text{ (calc)} = 79.81 \dfrac{\text{g}}{\text{mole}} \text{ (corr)}$

9.59　Compound A $= \dfrac{3.62 \text{ g}}{0.0521 \text{ mole}} = 69.48176583 \text{ (calc)} = 69.5 \dfrac{\text{g}}{\text{mole}} \text{ (corr)}$

　　　Compound B $= \dfrac{83.64 \text{ g}}{1.23 \text{ moles}} = 68 \text{ (calc)} = 68.0 \dfrac{\text{g}}{\text{mole}} \text{ (corr)}$

　　　Compound A has the greater molar mass.

RELATIONSHIP BETWEEN ATOMIC MASS UNITS AND GRAM UNITS (SEC. 9.8)

9.61　a)　$19.00 \text{ amu} \times \dfrac{1.000 \text{ g}}{6.022 \times 10^{23} \text{ amu}} = 3.155098 \times 10^{-23} \text{ (calc)} = 3.156 \times 10^{-23} \text{ g (corr)}$

　　　b)　$52.00 \text{ amu} \times \dfrac{1.000 \text{ g}}{6.022 \times 10^{23} \text{ amu}} = 8.635004982 \times 10^{-23} \text{ (calc)} = 8.635 \times 10^{-23} \text{ g (corr)}$

　　　c)　$118.71 \text{ amu} \times \dfrac{1.000 \text{ g}}{6.022 \times 10^{23} \text{ amu}} = 1.971272003 \times 10^{-22} \text{ (calc)} = 1.971 \times 10^{-23} \text{ g (corr)}$

　　　d)　$196.97 \text{ amu} \times \dfrac{1.000 \text{ g}}{6.022 \times 10^{23} \text{ amu}} = 3.270840252 \times 10^{-22} \text{ (calc)} = 3.271 \times 10^{-23} \text{ g (corr)}$

9.63　$5.143 \times 10^{-23} \text{ g} \times \dfrac{6.022 \times 10^{23} \text{ amu}}{1.000 \text{ g}} = 30.971146 \text{ (calc)} = 30.97 \text{ amu (corr)}$

The element with an atomic mass of 30.97 amu is phosphorus.

9.65　$26.98 \text{ g Al} \times \dfrac{6.022 \times 10^{23} \text{ amu}}{1.000 \text{ g}} \times \dfrac{1 \text{ atom Al}}{26.98 \text{ amu}} = 6.022 \times 10^{23} \text{ atoms Al (calc and corr)}$

THE MOLE AND CHEMICAL FORMULAS (SEC. 9.9)

9.67　$\dfrac{3 \text{ moles Na}}{1 \text{ mole Na}_3\text{PO}_4}$　　$\dfrac{1 \text{ mole P}}{1 \text{ mole Na}_3\text{PO}_4}$　　$\dfrac{4 \text{ moles O}}{1 \text{ mole Na}_3\text{PO}}$

　　　$\dfrac{3 \text{ moles Na}}{1 \text{ mole P}}$　　$\dfrac{3 \text{ moles Na}}{4 \text{ moles O}}$　　$\dfrac{1 \text{ mole P}}{4 \text{ moles O}}$

9.69　a)　$\dfrac{4 \text{ moles N}}{1 \text{ mole S}_4\text{N}_4\text{Cl}_2}$　b)　$\dfrac{2 \text{ moles Cl}}{1 \text{ mole S}_4\text{N}_4\text{Cl}_2}$　c)　$\dfrac{10 \text{ moles atoms}}{1 \text{ mole S}_4\text{N}_4\text{Cl}_2}$　d)　$\dfrac{2 \text{ moles Cl}}{4 \text{ moles S}}$

9.71　a)　$1.0 \text{ mole Na}_2\text{SO}_4 \times \dfrac{1 \text{ mole S}}{1 \text{ mole Na}_2\text{SO}_4} = 1.0 \text{ mole S (calc and corr)};$

　　　　$0.50 \text{ mole Na}_2\text{S}_2\text{O}_3 \times \dfrac{2 \text{ moles S}}{1 \text{ mole Na}_2\text{S}_2\text{O}_3} = 1.0 \text{ mole (calc and corr). Same moles of S.}$

　　　b)　$2.00 \text{ moles S}_3\text{Cl}_2 \times \dfrac{3 \text{ moles S}}{1 \text{ mole S}_3\text{Cl}_2} = 6.00 \text{ moles S (calc and corr)};$

　　　　$1.50 \text{ moles S}_2\text{O} \times \dfrac{2 \text{ moles S}}{1 \text{ mole S}_2\text{O}} = 3.00 \text{ moles (calc and corr). Not the same.}$

　　　c)　$3.00 \text{ moles H}_2\text{S}_2\text{O}_5 \times \dfrac{2 \text{ moles S}}{1 \text{ mole H}_2\text{S}_2\text{O}_5} = 6.00 \text{ moles S (calc and corr)};$

　　　　$6.00 \text{ moles H}_2\text{SO}_4 \times \dfrac{1 \text{ mole S}}{1 \text{ mole H}_2\text{SO}_4} = 6.00 \text{ moles S (calc and corr). Same moles of S.}$

　　　d)　$1.00 \text{ mole Na}_3\text{Ag(S}_2\text{O}_3)_2 \times \dfrac{4 \text{ moles S}}{1 \text{ mole Na}_3\text{Ag(S}_2\text{O}_3)_2} = 4.00 \text{ moles S (calc and corr)};$

　　　　$2.00 \text{ moles S}_2\text{F}_{10} \times \dfrac{2 \text{ moles S}}{1 \text{ mole S}_2\text{F}_{10}} = 4.00 \text{ moles S (calc and corr). Same moles of S.}$

THE MOLE AND CHEMICAL CALCULATIONS (SEC. 9.10)

9.73   a) $23.0 \text{ g Be} \times \dfrac{1 \text{ mole Be}}{9.01 \text{ g Be}} \times \dfrac{6.022 \times 10^{23} \text{ atoms Be}}{1 \text{ mole Be}}$

$$= 1.537247503 \times 10^{24} \text{ (calc)} = 1.54 \times 10^{24} \text{ atoms Be (corr)}$$

b) $23.0 \text{ g Mg} \times \dfrac{1 \text{ mole Mg}}{24.31 \text{ g Mg}} \times \dfrac{6.022 \times 10^{23} \text{ atoms Mg}}{1 \text{ mole Mg}}$

$$= 5.697490745 \times 10^{23} \text{ (calc)} = 5.70 \times 10^{23} \text{ atoms Mg (corr)}$$

c) $23.0 \text{ g Ca} \times \dfrac{1 \text{ mole Ca}}{48.08 \text{ g Ca}} \times \dfrac{6.022 \times 10^{23} \text{ atoms Ca}}{1 \text{ mole Ca}}$

$$= 3.455738523 \times 10^{23} \text{ (calc)} = 3.46 \times 10^{23} \text{ atoms Ca (corr)}$$

d) $23.0 \text{ g Sr} \times \dfrac{1 \text{ mole Sr}}{87.62 \text{ g Sr}} \times \dfrac{6.022 \times 10^{23} \text{ atoms Sr}}{1 \text{ mole Sr}}$

$$= 1.580757818 \times 10^{23} \text{ (calc)} = 1.58 \times 10^{23} \text{ atoms Sr (corr)}$$

9.75   a) $25.0 \text{ g NO} \times \dfrac{1 \text{ mole NO}}{30.01 \text{ g NO}} \times \dfrac{6.022 \times 10^{23} \text{ molecules NO}}{1 \text{ mole NO}}$

$$= 5.016661113 \times 10^{23} \text{ (calc)} = 5.02 \times 10^{23} \text{ molecules NO (corr)}$$

b) $25.0 \text{ g N}_2\text{O} \times \dfrac{1 \text{ mole N}_2\text{O}}{44.02 \text{ g N}_2\text{O}} \times \dfrac{6.022 \times 10^{23} \text{ molecules N}_2\text{O}}{1 \text{ mole N}_2\text{O}}$

$$= 3.420036347 \times 10^{23} \text{ (calc)} = 3.42 \times 10^{23} \text{ molecules N}_2\text{O (corr)}$$

c) $25.0 \text{ g N}_2\text{O}_3 \times \dfrac{1 \text{ mole N}_2\text{O}_3}{76.02 \text{ g N}_2\text{O}_3} \times \dfrac{6.022 \times 10^{23} \text{ molecules N}_2\text{O}_3}{1 \text{ mole N}_2\text{O}_3}$

$$= 1.980399895 \times 10^{23} \text{ (calc)} = 1.98 \times 10^{23} \text{ molecules N}_2\text{O}_3 \text{ (corr)}$$

d) $25.0 \text{ g N}_2\text{O}_5 \times \dfrac{1 \text{ mole N}_2\text{O}_5}{108.02 \text{ g N}_2\text{O}_5} \times \dfrac{6.022 \times 10^{23} \text{ molecules N}_2\text{O}_5}{1 \text{ mole N}_2\text{O}_5}$

$$= 1.393723385 \times 10^{23} \text{ (calc)} = 1.39 \times 10^{23} \text{ molecules N}_2\text{O}_5 \text{ (corr)}$$

9.77   a) $3.333 \times 10^{23} \text{ atoms P} \times \dfrac{1 \text{ mole P}}{6.022 \times 10^{23} \text{ atoms P}} \times \dfrac{30.97 \text{ g P}}{1 \text{ mole P}} = 17.14098472 \text{ (calc)}$

$$= 17.14 \text{ g P (corr)}$$

b) $3.333 \times 10^{23} \text{ molecules PH}_3 \times \dfrac{1 \text{ mole PH}_3}{6.022 \times 10^{23} \text{ atoms PH}_3} \times \dfrac{34.00 \text{ g PH}_3}{1 \text{ mole PH}_3} = 18.81800066 \text{ (calc)}$

$$= 18.82 \text{ g PH}_3 \text{ (corr)}$$

c) $2431 \text{ atoms P} \times \dfrac{1 \text{ mole P}}{6.022 \times 10^{23} \text{ molecules P}} \times \dfrac{30.97 \text{ g P}}{1 \text{ mole P}}$

$$= 1.250217038 \times 10^{-19} \text{ (calc)} = 1.250 \times 10^{-19} \text{ g P (corr)}$$

d) $2431 \text{ molecules PH}_3 \times \dfrac{1 \text{ mole PH}_3}{6.022 \times 10^{23} \text{ atoms PH}_3} \times \dfrac{34.00 \text{ g PH}_3}{1 \text{ mole PH}_3}$

$$= 1.372534042 \times 10^{-19} \text{ (calc)} = 1.373 \times 10^{-19} \text{ g PH}_3 \text{ (corr)}$$

9.79 a) $1 \text{ atom Na} \times \dfrac{22.99 \text{ amu}}{1 \text{ atom Na}} \times \dfrac{1 \text{ g}}{6.022 \times 10^{23} \text{ amu}} = 3.8176685 \times 10^{-23} \text{ (calc)}$

$$= 3.818 \times 10^{-23} \text{ g Na (corr)}$$

b) $1 \text{ atom Mg} \times \dfrac{24.31 \text{ amu}}{1 \text{ atom Mg}} \times \dfrac{1 \text{ g}}{6.022 \times 10^{23} \text{ amu}} = 4.0368648 \times 10^{-23} \text{ (calc)}$

$$= 4.037 \times 10^{-23} \text{ g Mg (corr)}$$

c) $1 \text{ molecule C}_4\text{H}_{10} \times \dfrac{[4(12.01) + 10(1.01)] \text{ amu}}{1 \text{ molecule C}_4\text{H}_{10}} \times \dfrac{1 \text{ g}}{6.022 \times 10^{23} \text{ amu}}$

$$= 9.6545998 \times 10^{-23} \text{ (calc)} = 9.655 \times 10^{-23} \text{ g C}_4\text{H}_{10} \text{ (corr)}$$

d) $1 \text{ molecule C}_6\text{H}_6 \times \dfrac{[6(12.01) + 6(1.01)] \text{ amu}}{1 \text{ molecule C}_6\text{H}_6} \times \dfrac{1 \text{ g}}{6.022 \times 10^{23} \text{ amu}}$

$$= 1.2972434 \times 10^{-22} \text{ (calc)} = 1.297 \times 10^{-22} \text{ g C}_6\text{H}_6 \text{ (corr)}$$

9.81 a) $100.0 \text{ g NaCl} \times \dfrac{1 \text{ mole NaCl}}{58.44 \text{ g NaCl}} \times \dfrac{1 \text{ mole Cl}}{1 \text{ mole NaCl}} \times \dfrac{35.45 \text{ g Cl}}{1 \text{ mole Cl}} = 60.660506 \text{ (calc)}$

$$= 60.66 \text{ g Cl (corr)}$$

b) $980.0 \text{ g CCl}_4 \times \dfrac{1 \text{ mole CCl}_4}{153.81 \text{ g CCl}_4} \times \dfrac{4 \text{ moles Cl}}{1 \text{ mole CCl}_4} \times \dfrac{35.45 \text{ g Cl}}{1 \text{ mole Cl}} = 903.47832 \text{ (calc)}$

$$= 903.5 \text{ g Cl (corr)}$$

c) $10.0 \text{ g HCl} \times \dfrac{1 \text{ mole HCl}}{36.46 \text{ g HCl}} \times \dfrac{1 \text{ mole Cl}}{1 \text{ mole HCl}} \times \dfrac{35.45 \text{ g Cl}}{1 \text{ mole Cl}} = 9.7229841 \text{ (calc)}$

$$= 9.72 \text{ g Cl (corr)}$$

d) $50.0 \text{ g BaCl}_2 \times \dfrac{1 \text{ mole BaCl}_2}{208.23 \text{ g BaCl}_2} \times \dfrac{2 \text{ moles Cl}}{1 \text{ mole BaCl}_2} \times \dfrac{35.45 \text{ g Cl}}{1 \text{ mole Cl}} = 17.024444 \text{ (calc)}$

$$= 17.0 \text{ g Cl (corr)}$$

9.83 a) $25.0 \text{ g PF}_3 \times \dfrac{1 \text{ mole PF}_3}{87.97 \text{ g PF}_3} \times \dfrac{1 \text{ mole P}}{1 \text{ mole PF}_3} \times \dfrac{6.022 \times 10^{23} \text{ atoms P}}{1 \text{ mole P}}$

$$= 1.7113789 \times 10^{23} \text{ (calc)} = 1.71 \times 10^{23} \text{ atoms P (corr)}$$

b) $25.0 \text{ g Be}_3\text{P}_2 \times \dfrac{1 \text{ mole Be}_3\text{P}_2}{88.97 \text{ g Be}_3\text{P}_2} \times \dfrac{2 \text{ moles P}}{1 \text{ mole Be}_3\text{P}_2} \times \dfrac{6.022 \times 10^{23} \text{ atoms P}}{1 \text{ mole P}}$

$$= 3.3842868 \times 10^{23} \text{ (calc)} = 3.38 \times 10^{23} \text{ atoms P (corr)}$$

c) $25.0 \text{ g POCl}_3 \times \dfrac{1 \text{ mole POCl}_3}{153.32 \text{ g POCl}_3} \times \dfrac{1 \text{ mole P}}{1 \text{ mole POCl}_3} \times \dfrac{6.022 \times 10^{23} \text{ atoms P}}{1 \text{ mole P}}$

$$= 9.8193321 \times 10^{22} \text{ (calc)} = 9.82 \times 10^{22} \text{ atoms P (corr)}$$

d) $25.0 \text{ g Na}_5\text{P}_3\text{O}_{10} \times \dfrac{1 \text{ mole Na}_5\text{P}_3\text{O}_{10}}{367.86 \text{ g Na}_5\text{P}_3\text{O}_{10}} \times \dfrac{3 \text{ moles P}}{1 \text{ mole Na}_5\text{P}_3\text{O}_{10}} \times \dfrac{6.022 \times 10^{23} \text{ atoms P}}{1 \text{ mole P}}$

$$= 1.2277769 \times 10^{23} \text{ (calc)} = 1.23 \times 10^{23} \text{ atoms P (corr)}$$

9.85   a) $4.7 \times 10^{24} \text{ molecules XeO}_3 \times \dfrac{1 \text{ mole XeO}_3}{6.022 \times 10^{23} \text{ molecules XeO}_3} \times \dfrac{3 \text{ moles O}}{1 \text{ mole XeO}_3} \times \dfrac{16.00 \text{ g O}}{1 \text{ mole O}}$

$$= 374.62637 \text{ (calc)} = 3.7 \times 10^2 \text{ g O (corr)}$$

b) $55.00 \text{ g SO}_2\text{Cl}_2 \times \dfrac{1 \text{ mole SO}_2\text{Cl}_2}{134.96 \text{ g SO}_2\text{Cl}_2} \times \dfrac{2 \text{ moles O}}{1 \text{ mole SO}_2\text{Cl}_2} \times \dfrac{16.00 \text{ g O}}{1 \text{ mole O}} = 13.04090101 \text{ (calc)}$

$$= 13.04 \text{ g O (corr)}$$

c) $0.30 \text{ moles C}_6\text{H}_{12}\text{O}_6 \times \dfrac{6 \text{ moles O}}{1 \text{ mole C}_6\text{H}_{12}\text{O}_6} \times \dfrac{16.00 \text{ g O}}{1 \text{ mole O}} = 28.8 \text{ (calc)} = 29 \text{ g O (corr)}$

d) $475 \text{ g Na}_2\text{CO}_3 \times \dfrac{1 \text{ mole Na}_2\text{CO}_3}{105.99 \text{ g Na}_2\text{CO}_3} \times \dfrac{3 \text{ moles O}}{1 \text{ mole Na}_2\text{CO}_3} \times \dfrac{16.00 \text{ g O}}{1 \text{ mole O}} = 215.1146335 \text{ (calc)}$

$$= 215 \text{ g O (corr)}$$

9.87   a) $14.84 \text{ g Al} \times \dfrac{1 \text{ mole Al}}{26.98 \text{ g Al}} = 0.550037064 \text{ (calc)} = 0.5500 \text{ mole Al (corr)}$

$14.84 \text{ g Al} \times \dfrac{1 \text{ mole Al}}{26.98 \text{ g Al}} \times \dfrac{6.022 \times 10^{23} \text{ atoms Al}}{1 \text{ mole Al}}$

$$= 3.312323202 \times 10^{23} \text{ (calc)} = 3.312 \times 10^{23} \text{ atoms Al (corr)}$$

b) $0.6379 \text{ mole Al} \times \dfrac{26.98 \text{ g Al}}{1 \text{ mole Al}} = 17.210542 \text{ (calc)} = 17.21 \text{ g Al (corr)}$

$0.6379 \text{ mole Al} \times \dfrac{6.022 \times 10^{23} \text{ atoms Al}}{1 \text{ mole Al}} = 3.8414338 \times 10^{23} \text{ (calc)}$

$$= 3.841 \times 10^{23} \text{ atoms Al (corr)}$$

c) $6.026 \times 10^{23} \text{ atoms Al} \times \dfrac{1 \text{ mole Al}}{6.022 \times 10^{23} \text{ atoms Al}} = 1.000664231 \text{ (calc)}$

$$= 1.001 \text{ moles Al (corr)}$$

$6.026 \times 10^{23} \text{ atoms Al} \times \dfrac{1 \text{ mole Al}}{6.022 \times 10^{23} \text{ atoms Al}} \times \dfrac{26.98 \text{ g Al}}{1 \text{ mole Al}}$

$$= 26.99792096 \text{ (calc)} = 27.00 \text{ g Al (corr)}$$

d) $1.384 \text{ moles Al} \times \dfrac{26.98 \text{ g Al}}{1 \text{ mole Al}} = 37.34032 \text{ (calc)} = 37.34 \text{ g Al (corr)}$

$1.384 \text{ moles Al} \times \dfrac{6.022 \times 10^{23} \text{ atoms Al}}{1 \text{ mole Al}} = 8.334448 \times 10^{23} \text{ (calc)}$

$= 8.334 \times 10^{23} \text{ atoms Al (corr)}$

	Grams of Sample	Moles of Sample	Atoms of Sample
	10.00	0.3706	2.232×10^{23}
a)	14.84	0.5500	3.312×10^{23}
b)	17.21	0.6379	3.841×10^{23}
c)	27.00	1.001	6.026×10^{23}
d)	37.34	1.384	8.334×10^{23}

9.89 a) $16.22 \text{ g CO}_2 \times \dfrac{1 \text{ mole CO}_2}{44.01 \text{ g CO}_2} = 0.368552601 \text{ (calc)}$

$= 0.3686 \text{ mole CO}_2 \text{ (corr)}$

$16.22 \text{ g CO}_2 \times \dfrac{1 \text{ mole CO}_2}{44.01 \text{ g CO}_2} \times \dfrac{6.022 \times 10^{23} \text{ molecules CO}_2}{1 \text{ mole CO}_2}$

$= 2.219423767 \times 10^{23} \text{ (calc)} = 2.219 \times 10^{23} \text{ molecules CO}_2 \text{ (corr)}$

$16.22 \text{ g CO}_2 \times \dfrac{1 \text{ mole CO}_2}{44.01 \text{ g CO}_2} \times \dfrac{6.022 \times 10^{23} \text{ molecules CO}_2}{1 \text{ mole CO}_2} \times \dfrac{3 \text{ atoms}}{1 \text{ molecule CO}_2}$

$= 6.658271302 \times 10^{23} \text{ (calc)} = 6.658 \times 10^{23} \text{ atoms in CO}_2 \text{ (corr)}$

b) $0.5555 \text{ mole CO}_2 \times \dfrac{44.01 \text{ g CO}_2}{1 \text{ mole CO}_2} = 24.447555 \text{ g CO}_2 \text{ (calc)}$

$= 24.45 \text{ g CO}_2 \text{ (corr)}$

$0.5555 \text{ mole CO}_2 \times \dfrac{6.022 \times 10^{23} \text{ molecules CO}_2}{1 \text{ mole CO}_2} = 3.345221 \times 10^{23} \text{ (calc)}$

$= 3.345 \times 10^{23} \text{ molecules CO}_2 \text{ (corr)}$

$0.5555 \text{ mole CO}_2 \times \dfrac{6.022 \times 10^{23} \text{ molecules CO}_2}{1 \text{ mole CO}_2} \times \dfrac{3 \text{ atoms}}{1 \text{ molecule CO}_2}$

$= 1.0035663 \times 10^{24} \text{ (calc)} = 1.004 \times 10^{24} \text{ atoms in CO}_2 \text{ (corr)}$

c) $4.000 \times 10^{23} \text{ molecules CO}_2 \times \dfrac{1 \text{ mole CO}_2}{6.022 \times 10^{23} \text{ molecules CO}_2}$

$= 0.664231152 \text{ (calc)} = 0.6642 \text{ mole CO}_2 \text{ (corr)}$

$4.000 \times 10^{23} \text{ molecules CO}_2 \times \dfrac{1 \text{ mole CO}_2}{6.022 \times 10^{23} \text{ molecules CO}_2} \times \dfrac{44.01 \text{ g CO}_2}{1 \text{ mole CO}_2}$

$= 29.23281302 \text{ (calc)} = 29.23 \text{ g CO}_2 \text{ (corr)}$

$4.000 \times 10^{23} \text{ molecules CO}_2 \times \dfrac{3 \text{ atoms}}{1 \text{ molecule CO}_2} = 1.200 \times 10^{24} \text{ atoms in CO}_2 \text{ (calc and corr)}$

d) 2.500×10^{25} atoms in $CO_2 \times \dfrac{1 \text{ molecule } CO_2}{3 \text{ atoms } CO_2} = 8.33333333 \times 10^{24}$ (calc)

$$= 8.333 \times 10^{24} \text{ molecules } CO_2 \text{ (corr)}$$

2.500×10^{25} atoms in $CO_2 \times \dfrac{1 \text{ molecule } CO_2}{3 \text{ atoms } CO_2} \times \dfrac{1 \text{ mole } CO_2}{6.022 \times 10^{23} \text{ molecules } CO_2}$

$$= 13.838149 \text{ (calc)} = 13.84 \text{ mole } CO_2 \text{ (corr)}$$

2.500×10^{25} atoms in $CO_2 \times \dfrac{1 \text{ molecule } CO_2}{3 \text{ atoms } CO_2} \times \dfrac{1 \text{ mole } CO_2}{6.022 \times 10^{23} \text{ molecules } CO_2} \times \dfrac{44.01 \text{ g } CO_2}{1 \text{ mole } CO_2}$

$$= 609.0169379 \text{ (calc)} = 609.0 \text{ g } CO_2 \text{ (corr)}$$

	Grams of Sample	Moles of Sample	Molecules in Sample	Atoms in Sample
	10.00	0.2272	1.368×10^{23}	4.105×10^{23}
a)	16.22	0.3686	2.219×10^{23}	6.658×10^{23}
b)	24.45	0.5555	3.345×10^{23}	1.004×10^{24}
c)	29.23	0.6642	4.000×10^{23}	1.200×10^{24}
d)	609.0	13.84	8.333×10^{24}	2.500×10^{25}

9.91   a) 2.70×10^{23} atoms $N \times \dfrac{1 \text{ mole } N}{6.022 \times 10^{23} \text{ atoms } N} \times \dfrac{1 \text{ mole } NH_3}{1 \text{ mole } N} \times \dfrac{17.04 \text{ g } NH_3}{1 \text{ mole } NH_3}$

$$= 7.639986715 \text{ (calc)} = 7.64 \text{ g } NH_3 \text{ (corr)}$$

b) 2.70×10^{23} atoms $H \times \dfrac{1 \text{ mole } H}{6.022 \times 10^{23} \text{ atoms } H} \times \dfrac{1 \text{ mole } NH_3}{3 \text{ moles } H} \times \dfrac{17.04 \text{ g } NH_3}{1 \text{ mole } NH_3}$

$$= 2.546662238 \text{ (calc)} = 2.55 \text{ g } NH_3 \text{ (corr)}$$

c) There are 4 atoms in 1 molecule of NH_3 (1 atom N + 3 atoms H)

2.70×10^{23} atoms total $\times \dfrac{1 \text{ molecule } NH_3}{4 \text{ atoms}} \times \dfrac{1 \text{ mole } NH_3}{6.022 \times 10^{23} \text{ molecules } NH_3} \times \dfrac{17.04 \text{ g } NH_3}{1 \text{ mole } NH_3}$

$$= 1.909996679 \text{ (calc)} = 1.91 \text{ g } NH_3 \text{ (corr)}$$

d) 2.70×10^{25} molecules $NH_3 \times \dfrac{1 \text{ mole } NH_3}{6.022 \times 10^{23} \text{ molecules } NH_3} \times \dfrac{17.04 \text{ g } NH_3}{1 \text{ mole } NH_3}$

$$= 7.639986715 \text{ (calc)} = 7.64 \text{ g } NH_3 \text{ (corr)}$$

9.93   The molar mass of $C_{21}H_{30}O_2$ is $21(12.01) + 30(1.01) + 2(16.00) = 314.51$ g/mole (corr)

a) 25.00 g $C_{21}H_{30}O_2 \times \dfrac{1 \text{ mole } C_{21}H_{30}O_2}{314.51 \text{ g } C_{21}H_{30}O_2} \times \dfrac{(21 + 30 + 2) \text{ moles atoms}}{1 \text{ mole } C_{21}H_{30}O_2}$

$$= 4.2129026 \text{ (calc)} = 4.213 \text{ moles atoms (corr)}$$

b) $25.00 \text{ g } C_{21}H_{30}O_2 \times \dfrac{1 \text{ mole } C_{21}H_{30}O_2}{314.51 \text{ g } C_{21}H_{30}O_2} \times \dfrac{21 \text{ moles C}}{1 \text{ mole } C_{21}H_{30}O_2} \times \dfrac{6.022 \times 10^{23} \text{ atoms C}}{1 \text{ mole C}}$

$= 1.0052304 \times 10^{24} \text{ (calc)} = 1.005 \times 10^{24} \text{ atoms C (corr)}$

c) $25.00 \text{ g } C_{21}H_{30}O_2 \times \dfrac{1 \text{ mole } C_{21}H_{30}O_2}{314.51 \text{ g } C_{21}H_{30}O_2} \times \dfrac{2 \text{ moles O}}{1 \text{ mole } C_{21}H_{30}O_2} \times \dfrac{16.00 \text{ g O}}{1 \text{ mole O}}$

$= 2.5436393 \text{ (calc)} = 2.544 \text{ g O (corr)}$

d) $25.00 \text{ g } C_{21}H_{30}O_2 \times \dfrac{1 \text{ mole } C_{21}H_{30}O_2}{314.51 \text{ g } C_{21}H_{30}O_2} \times \dfrac{6.022 \times 10^{23} \text{ molecules } C_{21}H_{30}O_2}{1 \text{ mole } C_{21}H_{30}O_2}$

$= 4.7868112 \times 10^{22} \text{ (calc)} = 4.787 \times 10^{22} \text{ molecules } C_{21}H_{30}O_2 \text{ (corr)}$

PURITY OF SAMPLES (SEC. 9.11)

9.95   a) $325 \text{ g sample } Fe_2S_3 \times \dfrac{95.4 \text{ g } Fe_2S_3}{100 \text{ g sample } Fe_2S_3} = 3.1005 \times 10^2 \text{ (calc)} = 3.10 \times 10^2 \text{ g } Fe_2S_3 \text{ (corr)}$

      b) $325 \text{ g sample } Fe_2S_3 - 310. \text{ g } Fe_2S_3 = 15 \text{ g impurities in sample (calc and corr)}$

9.97   Percent impurity $= \dfrac{0.23 \text{ g impurity}}{32.21 \text{ g sample } CaCO_3} \times 100 = 0.7140639553 \text{ (calc)} = 0.71\% \text{ impurity}$

9.99 $100 \text{ g sample} - 85.0 \text{ g of } Cu_2S = 15.0 \text{ g are impurities in sample}$

     $3.00 \text{ g impurities} \times \dfrac{100 \text{ g sample}}{15.0 \text{ g impurities}} = 20 \text{ (calc)} = 20.0 \text{ g sample (corr)}$

9.101 $35.00 \text{ g sample} \times \dfrac{92.35 \text{ g Cu}}{100 \text{ g sample}} \times \dfrac{1 \text{ mole Cu}}{63.55 \text{ g Cu}} \times \dfrac{6.022 \times 10^{23} \text{ atoms Cu}}{1 \text{ mole Cu}}$

$= 3.062881117 \times 10^{23} \text{ (calc)} = 3.063 \times 10^{23} \text{ atoms Cu (corr)}$

9.103 $25.00 \text{ g sample} \times \dfrac{64.0 \text{ g } Cr_2O_3}{100 \text{ g sample}} \times \dfrac{104.00 \text{ g Cr}}{152.00 \text{ g } Cr_2O_3} = 1.094737 \times 10^1 \text{ (calc)} = 10.9 \text{ g Cr (corr)}$

DETERMINATION OF EMPIRICAL FORMULAS (SEC. 9.12)

9.105   a) N_2O_4/NO_2, divide each subscript of N_2O_4 by 2 to give NO_2, pairing is correct.
      b) PF_3/PF, pairing is incorrect.
      c) C_2H_6/CH_3, divide each subscript of C_2H_6 by 2 to give CH_3, pairing is correct.
      d) H_2O_2/H_2O, pairing is incorrect.

9.107   a) N_2O, empirical formula is N_2O.
        NO, empirical formula is NO.
        NO_2, empirical formula is NO_2.
        N_2O_4, divide each subscript by 2, empirical formula is NO_2.
        NO_2 and N_2O_4 have same empirical formula, NO2.

      b) C_2H_2, empirical formula is CH.
        C_2H_4, empirical formula is CH_2.
        C_3H_6, empirical formula is CH_2.
        C_4H_{10}, empirical formula is C_2H_5.
        C_2H_4 and C_3H_6 have the same empirical formula, CH_2.

9.109 a) C_2H_2, has an empirical formula, CH, not a possible molecular formula for CH_2.
b) C_2H_4, empirical formula is CH_2, the molecular formula is possible for CH_2.
c) C_3H_6, empirical formula is CH_2, the molecular formula is possible for CH_2.
d) C_3H_8, empirical formula is C_3H_8, not a possible molecular formula for CH_2.

9.111 a) Molecular and empirical formula is $C_6H_8O_7$.
b) Molecular formula is $C_6H_{12}O_6$; empirical formula is CH_2O.

9.113 a) $58.91 \text{ g Na} \times \dfrac{1 \text{ mole Na}}{22.99 \text{ g Na}} = 2.5624184 \text{ (calc)} = 2.562 \text{ moles Na (corr)}$

$41.09 \text{ g S} \times \dfrac{1 \text{ mole S}}{32.06 \text{ g S}} = 1.281659389 \text{ (calc)} = 1.282 \text{ moles S (corr)}$

Na: $\dfrac{2.562}{1.282} = 1.998$ S: $\dfrac{1.282}{1.282} = 1.000$ Empirical formula is Na_2S.

b) $24.74 \text{ g K} \times \dfrac{1 \text{ mole K}}{39.10 \text{ g K}} = 0.63273657 \text{ (calc)} = 0.6327 \text{ mole K (corr)}$

$34.76 \text{ g Mn} \times \dfrac{1 \text{ mole Mn}}{54.94 \text{ g Mn}} = 0.6326902 \text{ (calc)} = 0.6327 \text{ mole Mn (corr)}$

$40.50 \text{ g O} \times \dfrac{1 \text{ mole O}}{16.00 \text{ g O}} = 2.53125 \text{ (calc)} = 2.531 \text{ moles O (corr)}$

K: $\dfrac{0.6327}{0.6327} = 1.000$ Mn: $\dfrac{0.6327}{0.6327} = 1.000$ O: $\dfrac{2.531}{0.6327} = 4.000$ Empirical formula is $KMnO_4$.

c) $2.06 \text{ g H} \times \dfrac{1 \text{ mole H}}{1.01 \text{ g H}} = 2.0396039 \text{ (calc)} = 2.04 \text{ moles H (corr)}$

$32.69 \text{ g S} \times \dfrac{1 \text{ mole S}}{32.06 \text{ g S}} = 1.019650655 \text{ (calc)} = 1.020 \text{ moles S (corr)}$

$65.25 \text{ g O} \times \dfrac{1 \text{ mole O}}{16.00 \text{ g O}} = 4.078125 \text{ (calc)} = 4.078 \text{ moles O (corr)}$

H: $\dfrac{2.04}{1.020} = 2.00$ S: $\dfrac{1.020}{1.020} = 1.000$ O: $\dfrac{4.078}{1.020} = 3.998$ Empirical formula is H_2SO_4.

d) $19.84 \text{ g C} \times \dfrac{1 \text{ mole C}}{12.01 \text{ g C}} = 1.6519567 \text{ (calc)} = 1.652 \text{ moles C (corr)}$

$2.50 \text{ g H} \times \dfrac{1 \text{ mole H}}{1.01 \text{ g H}} = 2.4752475 \text{ (calc)} = 2.48 \text{ moles H (corr)}$

$66.08 \text{ g O} \times \dfrac{1 \text{ mole O}}{16.00 \text{ g O}} = 4.13 \text{ (calc)} = 4.130 \text{ moles O (corr)}$

$11.57 \text{ g N} \times \dfrac{1 \text{ mole N}}{14.01 \text{ g N}} = 0.82583868 \text{ (calc)} = 0.8258 \text{ mole N (corr)}$

C: $\dfrac{1.652}{0.8258} = 2.000$ H: $\dfrac{2.48}{0.8258} = 3.00$ O: $\dfrac{4.130}{0.8258} = 5.001$ N: $\dfrac{0.8258}{0.8258} = 1.000$

Empirical formula is $C_2H_3O_5N$.

9.115 a) $1.00 \times 3 = 3.00$, $1.67 \times 3 = 5.01$, $\therefore$ 3 to 5
  b) $1.00 \times 2 = 2.00$, $1.50 \times 2 = 3.00$, $\therefore$ 2 to 3
  c) $2.00 \times 3 = 6.00$, $2.33 \times 3 = 6.99$, $\therefore$ 6 to 7
  d) $1.33 \times 3 = 3.99$, $2.33 \times 3 = 6.99$, $2.00 \times 3 = 6.00$, $\therefore$ 4 to 7 to 6

9.117 a) $43.64 \text{ g P} \times \dfrac{1 \text{ mole P}}{30.97 \text{ g P}} = 1.4091055 \text{ (calc)} = 1.409 \text{ moles P (corr)}$

$56.36 \text{ g O} \times \dfrac{1 \text{ mole O}}{16.00 \text{ g O}} = 3.5225 \text{ (calc)} = 3.522 \text{ moles O (corr)}$

P: $\dfrac{1.409}{1.409} = 1.000$ O: $\dfrac{3.502}{1.409} = 2.500$

O: $2.500 \times 2 = 5.000$ P: $1.000 \times 2 = 2.000$ Empirical formula is P_2O_5.

  b) $72.24 \text{ g Mg} \times \dfrac{1 \text{ mole Mg}}{24.30 \text{ g Mg}} = 2.9716166 \text{ (calc)} = 2.972 \text{ moles Mg (corr)}$

$27.76 \text{ g N} \times \dfrac{1 \text{ mole N}}{14.01 \text{ g N}} = 1.9814418 \text{ (calc)} = 1.981 \text{ moles N (corr)}$

Mg: $\dfrac{2.972}{1.981} = 1.500$ N: $\dfrac{1.981}{1.981} = 1.000$

Mg: $1.500 \times 2 = 3.000$ N: $1.000 \times 2 = 2.000$ Empirical formula is Mg_3N_2.

  c) $29.08 \text{ g Na} \times \dfrac{1 \text{ mole Na}}{22.99 \text{ g Na}} = 1.2648977 \text{ (calc)} = 1.265 \text{ moles Na (corr)}$

$40.56 \text{ g S} \times \dfrac{1 \text{ mole S}}{32.06 \text{ g S}} = 1.265127885 \text{ (calc)} = 1.265 \text{ moles S (corr)}$

$30.36 \text{ g O} \times \dfrac{1 \text{ mole O}}{16.00 \text{ g O}} = 1.8975 \text{ (calc)} = 1.898 \text{ moles O (corr)}$

Na: $\dfrac{1.265}{1.265} = 1.000$ S: $\dfrac{1.265}{1.265} = 1.000$ O: $\dfrac{1.898}{1.265} = 1.500$

Na: $1.000 \times 2 = 2.000$ S: $1.000 \times 2 = 2.000$ O: $1.500 \times 2 = 3.000$
Empirical formula is $Na_2S_2O_3$.

  d) $21.85 \text{ g Mg} \times \dfrac{1 \text{ mole Mg}}{24.31 \text{ g Mg}} = 0.89880708 \text{ (calc)} = 0.8988 \text{ mole Mg (corr)}$

$27.83 \text{ g P} \times \dfrac{1 \text{ mole P}}{30.97 \text{ g P}} = 0.89861155 \text{ (calc)} = 0.8986 \text{ mole P (corr)}$

$50.32 \text{ g O} \times \dfrac{1 \text{ mole O}}{16.00 \text{ g O}} = 3.145 \text{ moles O (calc and corr)}$

Mg: $\dfrac{0.8988}{0.8986} = 1.000$ P: $\dfrac{0.8986}{0.8986} = 1.000$ O: $\dfrac{3.145}{0.8986} = 3.500$

Mg: $1.000 \times 2 = 2.000$ P: $1.000 \times 2 = 2.000$ O: $3.500 \times 2 = 7.000$
Empirical formula is $Mg_2P_2O_7$.

9.119 $5.798 \text{ g C} \times \dfrac{1 \text{ mole C}}{12.01 \text{ g C}} = 0.48276436 \text{ (calc)} = 0.4828 \text{ mole C (corr)}$

$1.46 \text{ g H} \times \dfrac{1 \text{ mole H}}{1.01 \text{ g H}} = 1.4455445 \text{ (calc)} = 1.45 \text{ moles H (corr)}$

$7.740 \text{ g S} \times \dfrac{1 \text{ mole S}}{32.06 \text{ g S}} = 0.241422333 \text{ (calc)} = 0.2414 \text{ mole S (corr)}$

C: $\dfrac{0.4828}{0.2414} = 2.000$ H: $\dfrac{1.45}{0.2414} = 6.01$ S: $\dfrac{0.2414}{0.2414} = 1.000$ Empirical formula is C_2H_6S.

9.121   Mass of oxygen in compound $= 5.55 \text{ g} - 2.00 \text{ g} = 3.55 \text{ g oxygen}$

$2.00 \text{ g Be} \times \dfrac{1 \text{ mole Be}}{9.01 \text{ g Be}} = 0.22197558 \text{ (calc)} = 0.222 \text{ moles Be (corr)}$

$3.55 \text{ g O} \times \dfrac{1 \text{ mole O}}{16.00 \text{ g O}} = 0.221875 \text{ (calc)} = 0.222 \text{ moles O (corr)}$

Be: $\dfrac{0.222}{0.222} = 1.00$ O: $\dfrac{0.222}{0.222} = 1.00$ Empirical formula is BeO.

FORMULA DETERMINATION USING COMBUSTION ANALYSIS (SEC. 9.13)

9.123   Since this compound has only C and H, find moles C and moles H from data:

a) $0.338 \text{ g CO}_2 \times \dfrac{1 \text{ mole CO}_2}{44.01 \text{ g CO}_2} \times \dfrac{1 \text{ mole C}}{1 \text{ mole CO}_2} = 0.007680072 \text{ (calc)} = 0.00768 \text{ moles C (corr)}$

$0.277 \text{ g H}_2O \times \dfrac{1 \text{ mole H}_2O}{18.02 \text{ g H}_2O} \times \dfrac{2 \text{ moles H}}{1 \text{ mole H}_2O} = 0.0307436 \text{ (calc)} = .0307 \text{ moles H (corr)}$

H: $\dfrac{0.0307}{0.00768} = 4.00$ C: $\dfrac{0.00768}{0.00768} = 1.00$ Empirical formula is CH_4.

b) $0.303 \text{ g CO}_2 \times \dfrac{1 \text{ mole CO}_2}{44.01 \text{ g CO}_2} \times \dfrac{1 \text{ mole C}}{1 \text{ mole CO}_2} = 0.006884799 \text{ (calc)} = 0.00688 \text{ moles C (corr)}$

$0.0621 \text{ g H}_2O \times \dfrac{1 \text{ mole H}_2O}{18.02 \text{ g H}_2O} \times \dfrac{2 \text{ mole H}}{1 \text{ mole H}_2O} = 0.0068923 \text{ (calc)} = 0.00689 \text{ moles H (corr)}$

H: $\dfrac{0.00689}{0.00688} = 1.00$ C: $\dfrac{0.00688}{0.00688} = 1.00$ Empirical formula is CH.

c) $0.225 \text{ g CO}_2 \times \dfrac{1 \text{ mole CO}_2}{44.01 \text{ g CO}_2} \times \dfrac{1 \text{ mole C}}{1 \text{ mole CO}_2} = 0.00511247 \text{ (calc)} = 0.00511 \text{ moles C (corr)}$

$0.115 \text{ g H}_2O \times \dfrac{1 \text{ mole H}_2O}{18.02 \text{ g H}_2O} \times \dfrac{2 \text{ mole H}}{1 \text{ mole H}_2O} = 0.0127635 \text{ (calc)} = 0.0128 \text{ moles H (corr)}$

H: $\dfrac{0.0128}{0.00511} = 2.50$ C: $\dfrac{0.00511}{0.00511} = 1.00$; Multiplying by 2 gives the empirical formula C_2H_5.

d) $0.314 \text{ g } CO_2 \times \dfrac{1 \text{ mole } CO_2}{44.01 \text{ g } CO_2} \times \dfrac{1 \text{ mole C}}{1 \text{ mole } CO_2} = 0.00713474 \text{ (calc)} = 0.00713 \text{ mole C}$

$0.192 \text{ g } H_2O \times \dfrac{1 \text{ mole } H_2O}{18.02 \text{ g } H_2O} \times \dfrac{2 \text{ moles H}}{1 \text{ mole } H_2O} = 0.0213096 \text{ (calc)} = 0.0213 \text{ moles H}$

H: $\dfrac{0.0213}{0.00713} = 2.99$ 　　　 C: $\dfrac{0.00713}{0.00713} = 1.00$ 　　　　　　 Empirical formula is CH_3.

9.125 　 $2.328 \text{ mg } CO_2 \times \dfrac{12.01 \text{ mg C}}{44.01 \text{ mg } CO_2} = 0.6352937 \text{ (calc)} = 0.6353 \text{ mg C (corr)}$

$0.7429 \text{ mg sample} - 0.6353 \text{ mg C} = 0.1076 \text{ mg H}$

$0.6353 \text{ mg C} \times \dfrac{1 \text{ g C}}{10^3 \text{ mg C}} \times \dfrac{1 \text{ mole C}}{12.01 \text{ g C}} = 5.2897587 \times 10^{-5} \text{ (calc)} = 5.290 \times 10^{-5} \text{ mole C (corr)}$

$0.1076 \text{ mg H} \times \dfrac{1 \text{ g C}}{10^3 \text{ mg C}} \times \dfrac{1 \text{ mole H}}{1.01 \text{ g H}} = 1.0653465 \times 10^{-4} \text{ (calc)} = 1.065 \times 10^{-4} \text{ mole H (corr)}$

C: $\dfrac{5.290 \times 10^{-5}}{5.290 \times 10^{-5}} = 1.000$ 　　　 H: $\dfrac{1.065 \times 10^{-4}}{5.290 \times 10^{-5}} = 2.013$ 　　　 Empirical formula is CH_2.

9.127 　 The grams of O and the moles of C, H, and O are needed.

C: $10.05 \text{ g } CO_2 \times \dfrac{1 \text{ mole } CO_2}{44.01 \text{ g } CO_2} \times \dfrac{1 \text{ mole C}}{1 \text{ mole } CO_2} = 0.22835719 \text{ (calc)} = 0.2284 \text{ mole C (corr)}$

$0.2284 \text{ mole C} \times \dfrac{12.01 \text{ g C}}{1 \text{ mole C}} = 2.743084 \text{ (calc)} = 2.743 \text{ g C (corr)}$

H: $2.47 \text{ g } H_2O \times \dfrac{1 \text{ mole } H_2O}{18.02 \text{ g } H_2O} \times \dfrac{2 \text{ moles H}}{1 \text{ mole } H_2O} = 0.27413984 \text{ (calc)} = 0.274 \text{ mole H (corr)}$

$0.274 \text{ mole H} \times \dfrac{1.01 \text{ g H}}{1 \text{ mole H}} = 0.27674 \text{ (calc)} = 0.277 \text{ g H (corr)}$

O: $3.750 \text{ g compound} - 2.743 \text{ g C} - 0.277 \text{ g H} = 0.73 \text{ (calc)} = 0.730 \text{ g O (corr)}$

$0.730 \text{ g O} \times \dfrac{1 \text{ mole O}}{16.00 \text{ g O}} = 0.045625 \text{ (calc)} = 0.0456 \text{ mole O (corr)}$

C: $\dfrac{0.2284}{0.0456} = 5.01$ 　　　 H: $\dfrac{0.274}{0.0456} = 6.01$ 　　　 O: $\dfrac{0.0456}{0.0456} = 1.00$ 　 Empirical formula is C_5H_6O.

DETERMINATION OF MOLECULAR FORMULAS (SEC. 9.14)

9.129 　 a) formula mass

　　　 $CH_2 = 12.01 \text{ g/mole C} + 2(1.01) \text{ g/mole H} = 14.03 \text{ g/mole}$

　　　 $\dfrac{\text{molecular formula mass}}{\text{empirical formula mass}} = \dfrac{42.08 \text{ g/mole}}{14.03 \text{ g/mole}} = 2.9992872 \text{ (calc)} = 3$

　　　　　　　　　　　　　　　　　　 molecular formula $= (CH_2)_3 = C_3H_6$

　　　 b) formula mass

　　　 $NaS_2O_3 = 22.99 \text{ g/mole Na} + 2(32.06) \text{ g/mole S} + 3(16.00 \text{ g/mole O}) = 135.12 \text{ g/mole}$

　　　 $\dfrac{\text{molecular formula mass}}{\text{empirical formula mass}} = \dfrac{270.26 \text{ g/mole}}{135.12 \text{ g/mole}} = 2 \text{ (calc)} = 2$

　　　　　　　　　　　　　　　　　　 molecular formula $= (NaS_2O_3)_2 = Na_2S_4O_6$

c) formula mass

$C_3H_6O_2 = 3(12.01)$ g/mole C $+ 6(1.01)$ g/mole H $+ 2(16.00)$ g/mole O $= 74.09$ g/mole

empirical formula mass $=$ molecular formula mass

$$\text{molecular formula} = (C_3H_6O_2)_1 = C_3H_6O_2$$

d) formula mass

CHN $= 12.01$ g/mole C $+ 1.01$ g/mole H $+ 14.01$ g/mole N $= 27.02$ g/mole

$$\frac{\text{molecular formula mass}}{\text{empirical formula mass}} = \frac{135.15 \text{ g/mole}}{27.03 \text{ g/mole}} = 5(\text{calc}) = 5$$

$$\text{molecular formula} = (CHN)_5 = C_5H_5N_5$$

9.131 CH_2, molar mass $= 12.01$ g/mole $+ 2(1.01$ g/mole$) = 14.03$ g/mole

a) $14.03 \times 2 = 28.06$ g/mole

b) 3×14.03 g/mole $= 42.09$ g/mole

c) 5×14.03 g/mole $= 70.15$ g/mole

d) 8×14.03 g/mole $= 112.24$ g/mole

9.133 molar mass of CH $= 12.01$ g/mole $+ 1.01$ g/mole $= 13.02$ g/mole

a) $\dfrac{24.31 \text{ g/mole}}{13.02 \text{ g/mole}} = 1.867127496,$

not a possible molar mass for empirical formula CH.

b) $\dfrac{26.04 \text{ g/mole}}{13.02 \text{ g/mole}} = 2,$

yes, possible molar mass for empirical formula CH.

c) $\dfrac{39.06 \text{ g/mole}}{13.02 \text{ g/mole}} = 3,$

yes, possible molar mass for empirical formula CH.

d) $\dfrac{45.09 \text{ g/mole}}{13.02 \text{ g/mole}} = 3.463133641,$

not a possible molar mass for empirical formula CH.

9.135 a) molar mass $CH_2 = 12.01$ g/mole $+ 2(1.01$ g/mole$) = 14.03$ g/mole

$\dfrac{56.12 \text{ g/mole}}{14.03 \text{ g/mole}} = 4$, molecular formula is C_4H_8.

b) The empirical formula for C_2H_4, is CH_2.

c) The empirical formula for C_3H_6 is CH_2.

molar mass $C_3H_6 = 3(12.01$ g/mole$) + 6(1.01$ g/mole$) = 42.09$ g/mole

d) molar mass C_2H_2, $2(12.01$ g/mole$) + 2(1.01$ g/mole$) = 26.04$ g/mole

	Empirical Formula	Molar Mass (g/mole)	Molecular Formula
	C_3H_7	86.20	C_6H_{14}
a)	CH_2	56.12	C_4H_8
b)	CH_2	28.06	C_2H_4
c)	CH_2	42.09	C_3H_6
d)	CH	26.04	C_2H_2

9.137 a) To have 3 X atoms in our molecular formula, the empirical formula must be multiplied by 3. $(XY_3)_3 = X_3Y_9$

b) To have 3 Y atoms in our molecule, the empirical formula will be the same as the molecular formula.

c) The total number of atoms in one molecule of the empirical formula $= 1\,X + 3\,Y = 4$ atoms. To have 8 atoms in the molecular formula, we multiply the empirical formula by 2. $(XY_3)_2 = X_2Y_6$

d) The total number of atoms in one molecule of the empirical formula $= 1\,X + 3\,Y = 4$ atoms. Tetratomic is defined as 4 atoms in our molecular formula, thus the empirical formula is the same as the molecular formula.

9.139 a) For the molar mass of the molecular compound to equal twice the molar mass of the empirical formula, the molecular formula must be 2x that of the empirical formula: $(C_2H_3O)_2 = C_4H_6O_2$

b) One molecule of the empirical formula, C_2H_3O, contains $2\,C + 3\,H + 1\,O = 6$ atoms. To have 18 atoms in one molecule, we multiply the empirical formula by 3. $(C_2H_3O)_3 = C_6H_9O_3$

c) One molecule of the empirical formula, C_2H_3O, contains $2\,C + 1\,O = 3\,C$ and O atoms. To have 18 C and O atoms, we multiply our empirical formula by 6. $(C_2H_3O)_6 = C_{12}H_{18}O_6$

d) The mass of 0.010 mole of the empirical formula

$$0.010 \text{ mole of empirical formula} \times \frac{43.05 \text{ g empirical formula}}{1 \text{ mole empirical formula}} = 0.4305 \text{ g}$$

To determine the molecular formula, we divide the molar mass of 0.010 mole molecular formula by the molar mass of 0.010 mole of the empirical formula.

$$\text{The molecular formula} = \frac{0.86 \text{ g molecular formula}}{0.4305 \text{ g empirical formula}} = 2 \text{ (calc)} = 2$$

$$\text{Molecular formula} = (C_2H_3O)_2 = C_4H_6O_2$$

9.141 a) $100.0 \text{ g compound} \times \dfrac{70.57 \text{ g C}}{100 \text{ g compound}} \times \dfrac{1 \text{ mole C}}{12.01 \text{ g C}} = 5.8759367 \text{ (calc)}$

$$= 5.876 \text{ moles C (corr)}$$

$$100.0 \text{ g compound} \times \frac{5.93 \text{ g H}}{100 \text{ g compound}} \times \frac{1 \text{ mole H}}{1.01 \text{ g H}} = 5.87129 \text{ (calc)} = 5.87 \text{ moles H (corr)}$$

$$100.0 \text{ g compound} \times \frac{23.49 \text{ g O}}{100 \text{ g compound}} \times \frac{1 \text{ mole O}}{16.00 \text{ g O}} = 1.468125 \text{ (calc)}$$

$$= 1.468 \text{ moles O (corr)}$$

$C: \dfrac{5.876}{1.468} = 4.003 \quad H: \dfrac{5.87}{1.468} = 4.00 \text{ (corr)} \quad O: \dfrac{1.468}{1.468} = 1.00 \quad$ Empirical formula $= C_4H_4O$

Empirical formula mass $= 4(12.01) \text{ g/mole C} + 4(1.01) \text{ g/mole H} + 1(16.00) \text{ g/mole O}$

$$= 68.08 \text{ g/mole}$$

$$\frac{\text{molecular formula mass}}{\text{empirical formula mass}} = \frac{136 \text{ g/mole}}{68.08 \text{ g/mole}} = 1.9976498 \text{ (calc)} = 2$$

$$\text{Molecular formula} = (C_4H_4O)_2 = C_8H_8O_2$$

b) $1.00 \text{ mole compound} \times \dfrac{136 \text{ g compound}}{1 \text{ mole compound}} \times \dfrac{70.57 \text{ g C}}{100 \text{ g compound}} \times \dfrac{1 \text{ mole C}}{12.01 \text{ g C}}$

$$= 7.991274 \text{ (calc)} = 8.00 \text{ moles C (corr)}$$

$1.00 \text{ mole compound} \times \dfrac{136 \text{ g compound}}{1 \text{ mole compound}} \times \dfrac{5.93 \text{ g H}}{100 \text{ g compound}} \times \dfrac{1 \text{ mole H}}{1.01 \text{ g H}}$

$$= 7.9849505 \text{ (calc)} = 7.98 \text{ moles H (corr)}$$

$1.00 \text{ mole compound} \times \dfrac{136 \text{ g compound}}{1 \text{ mole compound}} \times \dfrac{23.49 \text{ g O}}{100 \text{ g compound}} \times \dfrac{1 \text{ mole O}}{16.00 \text{ g O}}$

$$= 1.99665 \text{ (calc)} = 2 \text{ moles O, and molecular formula is } C_8H_8O_2$$

9.143 a) $100.0 \text{ g compound} \times \dfrac{40.0 \text{ g C}}{100 \text{ g compound}} \times \dfrac{1 \text{ mole C}}{12.01 \text{ g C}} = 3.3305579 \text{ (calc)} = 3.33 \text{ moles C (corr)}$

$100.0 \text{ g compound} \times \dfrac{6.71 \text{ g H}}{100 \text{ g compound}} \times \dfrac{1 \text{ mole H}}{1.01 \text{ g H}} = 6.6435644 \text{ (calc)} = 6.64 \text{ moles H (corr)}$

$100.0 \text{ g compound} \times \dfrac{53.3 \text{ g O}}{100 \text{ g compound}} \times \dfrac{1 \text{ mole O}}{16.00 \text{ g O}} = 3.33125 \text{ (calc)} = 3.33 \text{ moles O (corr)}$

H: $\dfrac{6.64}{3.33} = 1.99399399 \text{ (calc)} = 2$ C: $\dfrac{3.33}{3.33} = 1 \text{ (calc)} = 1$ O: $\dfrac{3.33}{3.33} = 1.00 \text{ (calc)} = 1$

Empirical formula is CH_2O.

Empirical formula mass $= 12.01 \text{ amu C} + 2(1.01) \text{ amu H} + 16.00 \text{ amu O} = 30.03 \text{ amu}$

$\dfrac{\text{molecular formula mass}}{\text{empirical formula mass}} = \dfrac{90.0 \text{ amu}}{30.03 \text{ amu}} = 2.997003 \text{ (calc)} = 3$

Molecular formula $= (CH_2O)_3 = C_3H_6O_3$

b) $1.00 \text{ mole compound} \times \dfrac{90.0 \text{ g compound}}{1 \text{ mole compound}} \times \dfrac{40.0 \text{ g C}}{100 \text{ g compound}} \times \dfrac{1 \text{ mole C}}{12.01 \text{ g C}}$

$$= 2.9975021 \text{ (calc)} = 3.00 \text{ moles (corr)}$$

$1.00 \text{ mole compound} \times \dfrac{90.0 \text{ g compound}}{1 \text{ mole compound}} \times \dfrac{6.71 \text{ g H}}{100 \text{ g compound}} \times \dfrac{1 \text{ mole H}}{1.01 \text{ g H}}$

$$= 5.979201 \text{ (calc)} = 5.98 \text{ moles H (corr)}$$

$1.00 \text{ mole compound} \times \dfrac{90.0 \text{ g compound}}{1 \text{ mole compound}} \times \dfrac{53.3 \text{ g O}}{100 \text{ g compound}} \times \dfrac{1 \text{ mole O}}{16.00 \text{ g O}}$

$$= 2.99813 \text{ (calc)} = 3.00 \text{ moles O (corr)}$$

Molecular formula $= C_3H_6O_3$

Multi-Concept Problems

9.145 a) gold   b) sulfur   c) Cl_2 molecules   d) Ne

9.147 a) P_4 b) 1.00 mol Na   c) Cu   d) Be

9.149 Method 1: Molecular mass of K_2S = 2(39.10) + 32.06 = 110.26 amu

$$4.000 \text{ g } K_2S \times \frac{1 \text{ mole } K_2S}{110.26 \text{ g } K_2S} \times \frac{2 \text{ moles } K}{1 \text{ mole } K_2S} \times \frac{39.10 \text{ g } K}{1 \text{ mole } K} = 2.8366736 \text{ (calc)}$$
$$= 2.837 \text{ g } K \text{ (corr)}$$

$$4.000 \text{ g } K_2S \times \frac{1 \text{ mole } K_2S}{110.26 \text{ g } K_2S} \times \frac{1 \text{ mole } S}{1 \text{ mole } K_2S} \times \frac{32.06 \text{ g } S}{1 \text{ mole } S} = 1.163069109 \text{ (calc)}$$
$$= 1.163 \text{ g } S \text{ (corr)}$$

Method 2: $\%K$ in $K_2S = \dfrac{2(39.10) \text{ amu}}{110.26 \text{ amu}} \times 100 = 70.9232722 \text{ (calc)} = 70.92\% \text{ K (corr)}$

$$\%S = \frac{32.06 \text{ amu}}{110.26 \text{ amu}} \times 100 = 29.0767277 \text{ (calc)} = 29.08\% \text{ S (corr)}$$

$$4.000 \text{ g } K_2S \times \frac{70.92 \text{ g } K}{100 \text{ g } K_2S} = 2.8368 \text{ (calc)} = 2.837 \text{ g } K \text{ (corr)}$$

$$4.000 \text{ g } K_2S \times \frac{29.08 \text{ g } S}{100 \text{ g } K_2S} = 1.1632 \text{ (calc)} = 1.163 \text{ g } S \text{ (corr)}$$

9.151 3.50 moles of B contain the same number of atoms as 3.50 moles of Xe

$$3.50 \text{ moles B} \times \frac{10.81 \text{ g B}}{\text{mole B}} = 37.835 \text{ (calc)} = 37.8 \text{ g B (corr)}$$

9.153 From Al_2O_3, for every 2 atoms of Al, we have 3 atoms of O.

$$7.23 \times 10^{24} \text{ atoms Al} \times \frac{3 \text{ atoms O}}{2 \text{ atoms Al}} \times \frac{1 \text{ mole O}}{6.022 \times 10^{23} \text{ atoms O}} \times \frac{16.00 \text{ g O}}{1 \text{ mole O}}$$
$$= 288.1434739 \text{ (calc)} = 288 \text{ g O (corr)}$$

9.155 a) 3(12.01) amu C + 9(1.01) amu H + 28.09 amu Si + 35.45 amu Cl = 108.66 amu

$$\text{mass } \% \text{ H} = \frac{9.09 \text{ amu H}}{108.66 \text{ amu}} \times 100 = 8.3655439 \text{ (calc)} = 8.37\% \text{ H (corr)}$$

b) $1 \text{ molecule } (CH_3)_3SiCl \times \dfrac{14 \text{ atoms total}}{1 \text{ molecule } (CH_3)_3SiCl} = 14 \text{ atoms total (calc and corr)}$

$$1 \text{ molecule } (CH_3)_3SiCl \times \frac{9 \text{ H atoms}}{1 \text{ molecule } (CH_3)_3SiCl} = 9 \text{ H atoms (calc and corr)}$$

$$\text{atom } \% \text{ H} = \frac{9 \text{ H atoms}}{14 \text{ total atoms}} \times 100 = 64.285714 \text{ (calc)} = 64.29\% \text{ H (corr)}$$

c) $1 \text{ mole } (CH_3)_3SiCl \times \dfrac{14 \text{ moles atoms}}{1 \text{ mole } (CH_3)_3SiCl} = 14 \text{ moles atoms (calc and corr)}$

$$1 \text{ mole } (CH_3)_3SiCl \times \frac{9 \text{ moles H atoms}}{1 \text{ mole } (CH_3)_3 \text{ SiCl}} = 9 \text{ moles H atoms (calc and corr)}$$

$$\text{mole } \% \text{ H} = \frac{9 \text{ moles H atoms}}{14 \text{ total moles atoms}} \times 100 = 64.285714 \text{ (calc)} = 64.29\% \text{ H (corr)}$$

9.157 a) Since the ratio of atoms equals the ratio of moles of atoms, the atom ratio can be used instead of the mole ratio to determine the empirical formula.

$$Na: \frac{9.0 \times 10^{23}}{3.0 \times 10^{23}} = 3.0 \qquad Al: \frac{3.0 \times 10^{23}}{3.0 \times 10^{23}} = 1.0 \qquad F: \frac{1.8 \times 10^{24}}{3.0 \times 10^{23}} = 6.0$$

Empirical formula is Na_3AlF_6.

b) $3.2 \text{ g S} \times \dfrac{1 \text{ mole S}}{32.06 \text{ g S}} = 0.09981285 \text{ (calc)} = 0.10 \text{ mole S (corr)}$

$1.20 \times 10^{23} \text{ atoms O} \times \dfrac{1 \text{ mole O}}{6.022 \times 10^{23} \text{ atoms O}} = 0.1992693457 \text{ mole O (calc)}$

$= 0.199 \text{ mole O (corr)}$

$S: \dfrac{0.10}{0.10} = 1.0 \qquad O: \dfrac{0.199}{0.10} = 2.0$ \qquad\qquad Empirical formula is SO_2.

c) 0.36 mole Ba, 0.36 mole C $17.2 \text{ g O} \times \dfrac{1 \text{ mole O}}{16.00 \text{ g O}} = 1.075 \text{ (calc)} = 1.08 \text{ moles O (corr)}$

$Ba: \dfrac{0.36}{0.36} = 1.0 \qquad C: \dfrac{0.36}{0.36} = 1.0 \qquad O: \dfrac{1.08}{0.36} = 3.0$ \qquad Empirical formula is $BaCO_3$.

d) $1.81 \times 10^{23} \text{ atoms H} \times \dfrac{1 \text{ mole H}}{6.022 \times 10^{23} \text{ atoms H}} = 0.300564697 \text{ mole H (calc)}$

$= 0.301 \text{ mole H (corr)}$

$10.65 \text{ g Cl} \times \dfrac{1 \text{ mole}}{35.45 \text{ g Cl}} = 0.30042313 \text{ (calc)} = 0.3004 \text{ mole Cl (corr)}$

0.30 mole O (from problem)

$H: \dfrac{0.301}{0.30} = 1.0 \qquad O: \dfrac{0.30}{0.30} = 1.0 \qquad Cl: \dfrac{0.3004}{0.30} = 1.0$ \qquad Empirical formula is $HClO$.

9.159 $5.25 \text{ g compound} \times \dfrac{92.26 \text{ g C}}{100 \text{ g compound}} \times \dfrac{1 \text{ mole C}}{12.01 \text{ g C}} \times \dfrac{6.022 \times 10^{23} \text{ atoms C}}{1 \text{ mole C}}$

$= 2.4286811 \times 10^{23} \text{ (calc)} = 2.43 \times 10^{23} \text{ atoms C (corr)}$

9.161 Since the compound contains only C and H, the mass of C in 13.75 g CO_2 plus the mass of H in 11.25 g H_2O equals the mass of the compound.

$\text{mass C} = 13.75 \text{ g } CO_2 \times \dfrac{1 \text{ mole } CO_2}{44.01 \text{ g } CO_2} \times \dfrac{1 \text{ mole C}}{1 \text{ mole } CO_2} \times \dfrac{12.01 \text{ g C}}{1 \text{ mole C}}$

$= 3.75227221 \text{ (calc)} = 3.752 \text{ g C (corr)}$

$\text{mass H} = 11.25 \text{ g } H_2O \times \dfrac{1 \text{ mole } H_2O}{18.02 \text{ g } H_2O} \times \dfrac{2 \text{ moles H}}{1 \text{ mole } H_2O} \times \dfrac{1.01 \text{ g H}}{1 \text{ mole H}}$

$= 1.26109878 \text{ (calc)} = 1.261 \text{ g H (corr)}$

$\text{mass compound} = 3.752 \text{ g C} + 1.261 \text{ g H} = 5.013 \text{ g compound}$

9.163 Both the moles and grams of C, H, and S are needed.

Carbon: $6.60 \text{ g CO}_2 \times \dfrac{1 \text{ mole CO}_2}{44.01 \text{ g CO}_2} \times \dfrac{1 \text{ mole C}}{1 \text{ mole CO}_2} = 0.1499659 \text{ (calc)} = 0.150 \text{ mole C (corr)}$

$0.150 \text{ mole C} \times \dfrac{12.01 \text{ g C}}{1 \text{ mole C}} = 1.8015 \text{ (calc)} = 1.80 \text{ g C (corr)}$

Hydrogen: $5.41 \text{ g H}_2\text{O} \times \dfrac{1 \text{ mole H}_2\text{O}}{18.02 \text{ g H}_2\text{O}} \times \dfrac{2 \text{ moles H}}{1 \text{ mole H}_2\text{O}} = 0.600444 \text{ (calc)} = 0.600 \text{ mole H (corr)}$

$0.600 \text{ mole H} \times \dfrac{1.01 \text{ g H}}{1 \text{ mole H}} = 0.606 \text{ g H (calc and corr)}$

Sulfur: $9.61 \text{ g SO}_2 \times \dfrac{1 \text{ mole SO}_2}{64.06 \text{ g SO}_2} \times \dfrac{1 \text{ mole S}}{1 \text{ mole SO}_2} = 0.15001561 \text{ (calc)} = 0.150 \text{ mole S (corr)}$

$0.150 \text{ mole S} \times \dfrac{32.06 \text{ g S}}{1 \text{ mole S}} = 4.809 \text{ (calc)} = 4.81 \text{ g S (corr)}$

a) C: $\dfrac{0.150}{0.150} = 1.00$ H: $\dfrac{0.600}{0.150} = 4.00$ S: $\dfrac{0.150}{0.150} = 1.00$ Empirical formula is CH_4S.

b) Sample mass $= 1.80 \text{ g C} + 0.606 \text{ g H} + 4.81 \text{ g S} = 7.216 \text{ (calc)} = 7.22 \text{ g sample burned (corr)}$

9.165 Basis: 100.0 g sample: Since all the F is in NaF,

mass NaF $= 100.0 \text{ g sample} \times \dfrac{18.1 \text{ g F}}{100 \text{ g sample}} \times \dfrac{1 \text{ mole F}}{19.00 \text{ g F}} \times \dfrac{1 \text{ mole NaF}}{1 \text{ mole F}} \times \dfrac{41.99 \text{ g NaF}}{1 \text{ mole NaF}}$
$= 40.001 \text{ (calc)} = 40.0 \text{ g NaF (corr)}$

$\%\text{NaF} = \dfrac{40.0 \text{ g NaF}}{100.0 \text{ g sample}} \times 100 = 40.0\% \text{ NaF}$

Since all N atoms are in $NaNO_3$, mass $NaNO_3 =$

$100.0 \text{ g sample} \times \dfrac{6.60 \text{ g N}}{100 \text{ g sample}} \times \dfrac{1 \text{ mole N}}{14.01 \text{ g N}} \times \dfrac{1 \text{ mole NaNO}_3}{1 \text{ mole N}} \times \dfrac{85.00 \text{ g NaNO}_3}{1 \text{ mole NaNO}_3}$
$= 40.042827 \text{ (calc)} = 40.0 \text{ g NaNO}_3 \text{ (corr)}$

$\%\text{NaNO}_3 = \dfrac{40.0 \text{ g NaNO}_3}{100.0 \text{ g sample}} \times 100 = 40.0\% \text{ NaNO}_3$

Mass
$\text{Na}_2\text{SO}_4 = \text{mass sample} - \text{mass NaF} - \text{mass NaNO}_3 = 100.0 - 40.0 - 40.0 = 20.0 \text{ g Na}_2\text{SO}_4$

$\%\text{Na}_2\text{SO}_4 = \dfrac{20.0 \text{ g Na}_2\text{SO}_4}{100.0 \text{ g sample}} \times 100 = 20.0\% \text{ Na}_2\text{SO}_4$

9.167 $100 \text{ g mixture} - (5.000 \text{ g CO}_2 + 10.00 \text{ g N}_2\text{O}) = 85.00 \text{ g H}_2\text{O}.$
Mass of each element in mixture:

$5.000 \text{ g CO}_2 \times \dfrac{12.01 \text{ g C}}{44.01 \text{ g CO}_2} = 1.364462622 \text{ (calc)} = 1.364 \text{ g C (corr)}$

$5.000 \text{ g CO}_2 \times \dfrac{32.00 \text{ g O}}{44.01 \text{ g CO}_2} = 3.63553737 \text{ (calc)} = 3.636 \text{ g O (corr)}$

$10.00 \text{ g N}_2\text{O} \times \dfrac{28.02 \text{ g N}}{44.02 \text{ g N}_2\text{O}} = 6.365288505 \text{ (calc)} = 6.365 \text{ g N (corr)}$

$$10.00 \text{ g N}_2\text{O} \times \frac{16.00 \text{ g O}}{44.02 \text{ g N}_2\text{O}} = 3.634711495 \text{ (calc)} = 3.635 \text{ g O (corr)}$$

$$85.00 \text{ g H}_2\text{O} \times \frac{2.02 \text{ g H}}{18.02 \text{ g H}_2\text{O}} = 9.528301887 \text{ (calc)} = 9.528 \text{ g H (corr)}$$

$$85.00 \text{ g H}_2\text{O} \times \frac{16.00 \text{ g O}}{18.02 \text{ g H}_2\text{O}} = 75.47169811 \text{ (calc)} = 75.47 \text{ g O (corr)}$$

Percent C in mixture: $\dfrac{1.346 \text{ g C}}{100. \text{ g mixture}} \times 100 = 1.346\% \text{ C (calc and corr)}$

Percent N in mixture: $\dfrac{6.365 \text{ g N}}{100. \text{ g mixture}} \times 100 = 6.365\% \text{ N (calc and corr)}$

Percent H in mixture: $\dfrac{9.528 \text{ g H}}{100. \text{ g mixture}} \times 100 = 9.528\% \text{ H (calc and corr)}$

Percent O: $= 3.636 \text{ g (CO}_2) + 3.635 \text{ g (N}_2\text{O}) + 75.47 \text{ g (H}_2\text{O}) = 82.741 \text{ (calc)} = 82.74 \text{ g O (corr)}$

$\dfrac{82.47 \text{ g O}}{100. \text{ g mixture}} \times 100 = 82.47\% \text{ (calc and corr)}$

Confirming answer: $1.346\% \text{ C} + 6.365\% \text{ N} + 9.528\% \text{ H} + 82.74\% \text{ O} = 99.979 \, (100\%)$

9.169   28.9% of the formula mass must be from the chlorine atom. $0.289 \times$ (formula mass) $= 35.45$ amu

formula mass $= 122.66435$ amu (calc) $= 123$ amu (corr)

formula mass $=$ at. mass K $+$ at. mass Cl $+$ X(at. mass O)

123 amu $\quad= 39.10$ amu $+ 35.45$ amu $+$ X(16.00 amu)

3.028125 $\quad\;\;= $ X. The value of X is the integer 3.

9.171   mass O $=$ total mass $-$ mass M $= 10.498 \text{ g} - 7.503 \text{ g} = 2.995 \text{ g O (calc and corr)}$

$$2.995 \text{ g O} \times \frac{1 \text{ mole O}}{16.00 \text{ g O}} \times \frac{1 \text{ mole M}}{1 \text{ mole O}} = 0.1871875 \text{ (calc)} = 0.1872 \text{ mole M (corr)}$$

molar mass $=$ g of M divided by the moles of M. $\dfrac{7.503 \text{ g}}{0.1872 \text{ mole}}$

$$= 40.08013 \text{ (calc)} = 40.08 \frac{\text{g}}{\text{mole}} \text{ (corr)}$$

Answers to Multiple-Choice Practice Test

MC 9.1	e	MC 9.2	c	MC 9.3	d	MC 9.4	c	MC 9.5	c
MC 9.6	b	MC 9.7	b	MC 9.8	d	MC 9.9	d	MC 9.10	d
MC 9.11	d	MC 9.12	c	MC 9.13	a	MC 9.14	c	MC 9.15	d
MC 9.16	e	MC 9.17	d	MC 9.18	d	MC 9.19	e	MC 9.20	a

CHAPTER TEN
Chemical Calculations Involving Chemical Equations

THE LAW OF CONSERVATION OF MASS (SEC. 10.1)

10.1 The total mass of products = total mass of reactants
   a) 127.10 g Cu + 34.00 g O_2 = 161.10 g reactants, 159.10 g CuO products (calc and corr)
   This is inconsistent with the law of conservation of mass.
   b) 76.15 g CS_2 + 96.00 g O_2 = 172.15 g reactants (calc and corr)
   44.01 g CO_2 + 128.14 g SO_2 = 172.15 g products (calc and corr)
   This is consistent with the law of conservation of mass.

10.3 The total mass of products = total mass of reactants
   $(16.05 \text{ g CH}_4) + (64.00 \text{ g O}_2) = (x \text{ g CO}_2) + (36.04 \text{ g H}_2\text{O})$
   Mass of reactants = $(16.05 \text{ g CH}_4) + (64.00 \text{ g O}_2)$ = 80.05 g reactants (calc and corr)
   Mass of products = $(x \text{ g CO}_2) + (36.04 \text{ g H}_2\text{O})$
   $(x \text{ g CO}_2)$ = 80.05 g reactants − $(36.04 \text{ g H}_2\text{O})$ = 44.01 g CO_2 (calc and corr)

10.5 The law of conservation of mass requires the mass of the reactants to equal the mass of the products.
   Mass of reactants:
   4.2 g sodium hydrogen carbonate + 10.0 g acetic acid = 14.2 g reactants (calc and corr)
   Mass of contents in vessel = 12.0 g (products)
   Mass of missing carbon dioxide = $(14.2 - 12.0)$ g = 2.2 g carbon dioxide (calc and corr)

10.7 Diagrams I and III are consistent with the law of conservation of mass.
   Diagram I = 4 units of 2 empty paired circles (8) + 6 solid unpaired circles = 14 total
   Diagram III = 6 units with a solid and empty circle (12) + 2 extra paired solid circles = 14 total
   The type and number of circles remain the same in III.

CHEMICAL EQUATION NOTATION (SECS. 10.2 AND 10.5)

10.9

	Reaction	Reactants	Products
a)	FeO + CO → Fe + CO_2	2 reactants, FeO, CO	2 products, Fe, CO_2
b)	$CaCO_3$ + H_2SO_4 → $CaSO_4$ + CO_2 + H_2O	2 reactants, $CaCO_3$, H_2SO_4	3 products, $CaSO_4$, CO_2, H_2O
c)	BaS → Ba + S	1 reactant, BaS	2 products, Ba, S
d)	S + O_2 → SO_2	2 reactants, S, O_2	1 product, SO_2

10.11   Answers "a," "b," and "c" contain symbols correctly written for gases in an equation and therefore are appropriate. However, in "d" Br is not correct; it should be Br_2, and therefore, this is not appropriate.

10.13   a) (s) means solid, (l) means liquid, (aq) means water (aqueous) solution, and (g) means gas.
    b) (s) means solid, (aq) means water (aqueous) solution, (s) means solid, (g) means gas, and (l) means liquid.

BALANCING CHEMICAL EQUATIONS (SEC. 10.4)

10.15   a) balanced   b) unbalanced   c) balanced   d) unbalanced

10.17

	Reaction	Atoms, Reactants	Atoms, Products
a)	$Xe + 2 F_2 \rightarrow XeF_4$	1 Xe, 4 F	1 Xe, 4 F
b)	$2 Sb + 3 Cl_2 \rightarrow 2 SbCl_3$	2 Sb, 6 Cl	2 Sb, 6 Cl
c)	$K_2SO_4 + BaCl_2 \rightarrow$ $BaSO_4 + 2 KCl$	2 K, 1 S, 4 O, 1 Ba, 2 Cl	1 Ba, 1 S, 4 O, 2 K, 2 Cl
d)	$4 NO_2 + 2 H_2O + O_2 \rightarrow$ $4 HNO_3$	4 N, 12 O, 4 H	4 H, 4 N, 12 O

10.19   a) $4 Fe + 3 O_2 \rightarrow 2 Fe_2O_3$
    b) $2 Al + 6 HCl \rightarrow 2 AlCl_3 + 3 H_2$
    c) $SiH_4 + 4 F_2 \rightarrow SiF_4 + 4 HF$
    d) $2 NaOH + H_2S \rightarrow Na_2S + 2 H_2O$

10.21   a) $3 PbO + 2 NH_3 \rightarrow 3 Pb + N_2 + 3 H_2O$
    b) $2 NaHCO_3 + H_2SO_4 \rightarrow Na_2SO_4 + 2 CO_2 + 2 H_2O$
    c) $TiO_2 + C + 2 Cl_2 \rightarrow TiCl_4 + CO_2$
    d) $2 NBr_3 + 3 NaOH \rightarrow N_2 + 3 NaBr + 3 HBrO$

10.23   a) $C_5H_{12} + 8 O_2 \rightarrow 5 CO_2 + 6 H_2O$ b) $2 C_5H_{10} + 15 O_2 \rightarrow 10 CO_2 + 10 H_2O$
    c) $C_5H_8 + 7 O_2 \rightarrow 5 CO_2 + 4 H_2O$ d) $C_5H_{10}O + 7 O_2 \rightarrow 5 CO_2 + 5 H_2O$

10.25   a) $Ca(OH)_2 + 2 HNO_3 \rightarrow Ca(NO_3)_2 + 2 H_2O$
    b) $BaCl_2 + (NH_4)_2SO_4 \rightarrow BaSO_4 + 2 NH_4Cl$
    c) $2 Fe(OH)_3 + 3 H_2SO_4 \rightarrow Fe_2(SO_4)_3 + 6 H_2O$
    d) $Na_3PO_4 + 3 AgNO_3 \rightarrow 3 NaNO_3 + Ag_3PO_4$

10.27   a) divide each coefficient by 3: $AgNO_3 + KCl \rightarrow AgCl + KNO_3$
    b) divide each coefficient by 2: $CS_2 + 3 O_2 \rightarrow CO_2 + 2 SO_2$
    c) multiply each coefficient by 2: $2 H_2 + O_2 \rightarrow 2 H_2O$
    d) multiply each coefficient by 2: $2 Ag_2CO_3 \rightarrow 4 Ag + 2 CO_2 + O_2$

10.29   Balance the chemical equation
            $Co + Cl_2 \rightarrow CoCl_3$
    by using coefficients in front of each chemical. This gives:
            $2 Co + 3 Cl_2 \rightarrow 2 CoCl_3$, product is cobalt(III) chloride.
    The student answer of
            $Co + Cl_2 \rightarrow CoCl_2$
    incorrectly changes the identity of the product from $CoCl_3$, cobalt(III) chloride, to $CoCl_2$, cobalt(II) chloride.

10.31   a) $6 A_2 + 2 B_2 \rightarrow 4 A_3 B$ b) $3 A_2 + 3 B_2 \rightarrow 6 AB$

10.33   Diagram III is consistent with Diagram I.

CLASSES OF CHEMICAL REACTIONS (SEC. 10.6)

10.35   a) synthesis   b) synthesis   c) double-replacement   d) single-replacement

10.37   a) The correct formula for zinc nitrate is $Zn(NO_3)_2$ and the symbol for copper is Cu;
   $Zn + Cu(NO_3)_2 \rightarrow Zn(NO_3)_2 + Cu$
   b) The correct formula for calcium oxide is CaO; $2 Ca + O_2 \rightarrow 2 CaO$.
   c) The correct formula for barium sulfate is $BaSO_4$, and for potassium nitrate it is
   KNO_3; $K_2SO_4 + Ba(NO_3)_2 \rightarrow BaSO_4 + 2 KNO_3$.
   d) The correct formula for oxygen is O_2, the symbol for silver is Ag; $2 Ag_2O \rightarrow 4 Ag + O_2$.

10.39   a) $Rb_2CO_3 \rightarrow Rb_2O + CO_2$ b) $SrCO_3 \rightarrow SrO + CO_2$
   c) $Al_2(CO_3)_3 \rightarrow Al_2O_3 + 3 CO_2$ d) $Cu_2CO_3 \rightarrow Cu_2O + CO_2$

10.41   Combustion reactions need O_2 as a reactant and produce CO_2 and H_2O.
   a) $2 C_4H_{10} + 13 O_2 \rightarrow 8 CO_2 + 10 H_2O$ b) $C_7H_{12} + 10 O_2 \rightarrow 7 CO_2 + 6 H_2O$
   c) $2 CH_4O + 3 O_2 \rightarrow 2 CO_2 + 4 H_2O$ d) $C_3H_8O_2 + 4 O_2 \rightarrow 3 CO_2 + 4 H_2O$

10.43   Determine the molecular ratios between CO_2 and H_2O in the following combustion reactions:

   a) $2 C_5H_{10} + 15 O_2 \rightarrow 10 CO_2 + 10 H_2O$; $\dfrac{10 CO_2}{10 H_2O} = 1 CO_2 : 1 H_2O$

   b) $2 C_8H_{10} + 21 O_2 \rightarrow 16 CO_2 + 10 H_2O$; $\dfrac{16 CO_2}{10 H_2O} = 8 CO_2 : 5 H_2O$

   c) $C_2H_6O + 3 O_2 \rightarrow 2 CO_2 + 3 H_2O$; $\dfrac{2 CO_2}{3 H_2O} = 2 CO_2 : 3 H_2O$

   d) $2 C_6H_{14}O_2 + 17 O_2 \rightarrow 12 CO_2 + 14 H_2O$; $\dfrac{12 CO_2}{14 H_2O} = 6 CO_2 : 7 H_2O$

10.45   a) $4 C_2H_7N + 19 O_2 \rightarrow 8 CO_2 + 14 H_2O + 4 NO_2$
   b) $CH_4S + 3 O_2 \rightarrow CO_2 + 2 H_2O + SO_2$

10.47   a) single-replacement, combustion, and synthesis
   b) single-replacement and decomposition
   c) synthesis, decomposition, single-replacement, double-replacement, and combustion
   d) synthesis, decomposition, single-replacement, double-replacement, and combustion

10.49   Reaction: $2 NH_3 \rightarrow 3 H_2 + N_2$

   Mole-to-mole conversion factors: $\dfrac{2 \text{ moles } NH_3}{3 \text{ moles } H_2}$ $\dfrac{2 \text{ moles } NH_3}{1 \text{ mole } N_2}$ $\dfrac{3 \text{ moles } H_2}{1 \text{ mole } N_2}$

   $\dfrac{3 \text{ moles } H_2}{2 \text{ moles } NH_3}$ $\dfrac{1 \text{ mole } N_2}{2 \text{ moles } NH_3}$ $\dfrac{1 \text{ mole } N_2}{3 \text{ moles } H_2}$

10.51 $6 ClO_2 + 3 H_2O \rightarrow 5 HClO_3 + HCl$

   a) $\dfrac{3 \text{ moles } H_2O}{6 \text{ moles } ClO_2}$ b) $\dfrac{1 \text{ mole } HCl}{5 \text{ moles } HClO_3}$ c) $\dfrac{1 \text{ mole } HCl}{3 \text{ moles } H_2O}$ d) $\dfrac{6 \text{ moles } ClO_2}{5 \text{ moles } HClO_3}$

10.53 a) $3.00 \text{ moles N}_2 \times \dfrac{2 \text{ moles NaN}_3}{3 \text{ moles N}_2} = 2 \text{ (calc)} = 2.00 \text{ moles NaN}_3 \text{ (corr)}$

b) $3.00 \text{ moles N}_2 \times \dfrac{3 \text{ moles CO}}{1 \text{ mole N}_2} = 9 \text{ (calc)} = 9.00 \text{ moles CO (corr)}$

c) $3.00 \text{ moles N}_2 \times \dfrac{2 \text{ moles NH}_2\text{Cl}}{1 \text{ mole N}_2} = 6 \text{ (calc)} = 6.00 \text{ moles NH}_2\text{Cl (corr)}$

d) $3.00 \text{ moles N}_2 \times \dfrac{4 \text{ moles C}_3\text{H}_5\text{O}_9\text{N}_3}{6 \text{ moles N}_2} = 2 \text{ (calc)} = 2.00 \text{ moles C}_3\text{H}_5\text{O}_9\text{N}_3 \text{ (corr)}$

10.55 a) $1.42 \text{ moles H}_2\text{S} \times \dfrac{1 \text{ mole H}_2\text{O}_2}{1 \text{ mole H}_2\text{S}} = 1.42 \text{ mole H}_2\text{O}_2 \text{ (calc and corr)}$

b) $1.42 \text{ moles O}_2 \times \dfrac{1 \text{ mole CS}_2}{3 \text{ moles O}_2} = 0.4733333 \text{ (calc)} = 0.473 \text{ mole CS}_2 \text{ (corr)}$

c) $1.42 \text{ moles HCl} \times \dfrac{1 \text{ mole Mg}}{2 \text{ moles HCl}} = 0.71 \text{ (corr)} = 0.710 \text{ mole Mg (corr)}$

d) $1.42 \text{ moles Al} \times \dfrac{6 \text{ moles HCl}}{2 \text{ moles Al}} = 4.26 \text{ moles HCl (calc and corr)}$

10.57 a) $1.75 \text{ moles NH}_4\text{NO}_3 \times \dfrac{7 \text{ moles products}}{2 \text{ moles NH}_4\text{NO}_3} = 6.125 \text{ (calc)} = 6.12 \text{ moles products (corr)}$

b) $1.75 \text{ moles NaClO}_3 \times \dfrac{5 \text{ moles products}}{2 \text{ moles NaClO}_3} = 4.375 \text{ (calc)} = 4.38 \text{ moles products (corr)}$

c) $1.75 \text{ moles KNO}_3 \times \dfrac{3 \text{ moles products}}{2 \text{ moles KNO}_3} = 2.625 \text{ (calc)} = 2.62 \text{ moles products (corr)}$

d) $1.75 \text{ moles I}_4\text{O}_9 \times \dfrac{11 \text{ moles products}}{4 \text{ moles I}_4\text{O}_9} = 4.8125 \text{ (calc)} = 4.81 \text{ moles products (corr)}$

10.59 a) $6.0 \text{ moles Ag} \times \dfrac{2 \text{ moles Ag}_2\text{CO}_3}{4 \text{ moles Ag}} = 3 \text{ (calc)} = 3.0 \text{ moles Ag}_2\text{CO}_3 \text{ (corr)}$

b) $8.0 \text{ moles CO}_2 \times \dfrac{2 \text{ moles Ag}_2\text{CO}_3}{2 \text{ moles CO}_2} = 8 \text{ (calc)} = 8.0 \text{ moles Ag}_2\text{CO}_3 \text{ (corr)}$

c) $2.4 \text{ moles O}_2 \times \dfrac{2 \text{ moles Ag}_2\text{CO}_3}{1 \text{ mole O}_2} = 4.8 \text{ moles Ag}_2\text{CO}_3 \text{ (calc and corr)}$

d) $15.0 \text{ moles of products} \times \dfrac{2 \text{ moles Ag}_2\text{CO}_3}{7 \text{ moles product}} = 4.2857142857 \text{ (calc)} = 4.29 \text{ moles Ag}_2\text{CO}_3 \text{ (corr)}$

BALANCED CHEMICAL EQUATIONS AND THE LAW OF CONSERVATION OF MASS (SEC. 10.8)

10.61 Law of Conservation of Mass states mass reactants = mass products
a) $4.00 \text{ g A} + 1.67 \text{ g B} = 5.67 \text{ g C}$
b) $3.76 \text{ g A} + x \text{ g B} = 7.02 \text{ g C}$ $\qquad\qquad \therefore 7.02 \text{ g C} - 3.76 \text{ g A} = 3.26 \text{ g B}$

10.63 a) $C_7H_{16} + 11 O_2 \rightarrow 7 CO_2 + 8 H_2O$
$100.23 \text{ g C}_7\text{H}_{16} + 11(32.00) \text{ g O}_2 = 7(44.01) \text{ g CO}_2 + 8(18.02) \text{ g H}_2\text{O}$
$452.23 \text{ g reactants} = 452.23 \text{ g products}$

b) $2\,HCl + CaCO_3 \rightarrow CaCl_2 + CO_2 + H_2O$
   $2(36.46)\,g\,HCl + 100.09\,g\,CaCO_3 = 110.98\,g\,CaCl_2 + 44.01\,g\,CO_2 + 18.02\,g\,H_2O$

$$173.01\,g\,reactants = 173.01\,g\,products$$

c) $4\,Na_2SO_4 + 2\,C \rightarrow Na_2S + 2\,CO_2$
   $142.04\,g\,Na_2SO_4 + 2(12.01)\,g\,C = 78.08\,g\,Na_2S + 2(44.01)\,g\,CO_2$

$$166.06\,g\,reactants = 166.06\,g\,products$$

d) $4\,Na_2CO_3 + Fe_3Br_8 \rightarrow 8\,NaBr + 4\,CO_2 + Fe_3O_4$
   $4(105.99)\,g\,Na_2CO_3 + 806.75\,g\,Fe_3Br_8 = 8(102.89)\,g\,NaBr + 4(44.01)\,g\,CO_2 + 231.55\,g\,Fe_3O_4$

$$1230.71\,g\,reactants = 1230.71\,g\,products$$

CALCULATIONS BASED ON CHEMICAL EQUATIONS (SEC. 10.9)

10.65   a) $2\,KClO_3 \rightarrow 2\,KCl + 3\,O_2$

$$2.68\,moles\,KClO_3 \times \frac{3\,moles\,O_2}{2\,moles\,KClO_3} = 4.02\,moles\,O_2\,(calc\ and\ corr)$$

b) $2\,CuO \rightarrow 2\,Cu + O_2$

$$2.68\,moles\,CuO \times \frac{1\,mole\,O_2}{2\,moles\,CuO} = 1.34\,moles\,O_2\,(calc\ and\ corr)$$

c) $2\,NaNO_3 \rightarrow 2\,NaNO_2 + O_2$

$$2.68\,moles\,NaNO_3 \times \frac{1\,mole\,O_2}{2\,moles\,NaNO_3} = 1.34\,moles\,O_2\,(calc\ and\ corr)$$

d) $4\,HNO_3 \rightarrow 4\,NO_2 + 2\,H_2O + O_2$

$$2.68\,moles\,HNO_3 \times \frac{1\,mole\,O_2}{4\,moles\,HNO_3} = 0.67\,(corr) = 0.670\,mole\,O_2\,(calc)$$

10.67   a) $CO + 3\,H_2 \rightarrow CH_4 + H_2O$

$$1.25\,moles\,H_2 \times \frac{1\,mole\,CO}{3\,moles\,H_2} = 0.4166667\,(calc) = 0.417\,mole\,CO\,(corr)$$

b) $CaO + 2\,HCl \rightarrow CaCl_2 + H_2O$

$$1.25\,moles\,HCl \times \frac{1\,mole\,CaO}{2\,moles\,HCl} = 0.625\,mole\,CaO\,(calc\ and\ corr)$$

c) $4\,NH_3 + 3\,O_2 \rightarrow 2\,N_2 + 6\,H_2O$

$$1.25\,moles\,O_2 \times \frac{4\,moles\,NH_3}{3\,moles\,O_2} = 1.6666667\,(calc) = 1.67\,moles\,NH_3\,(corr)$$

d) $2\,HCl + Ba(OH)_2 \rightarrow BaCl_2 + 2\,H_2O$

$$1.25\,moles\,Ba(OH)_2 \times \frac{2\,moles\,HCl}{1\,mole\,Ba(OH)_2} = 2.5\,(calc) = 2.50\,moles\,HCl\,(corr)$$

10.69   a) $CO + 3\,H_2 \rightarrow CH_4 + H_2O$

$$5.25\,moles\,H_2O \times \frac{1\,mole\,CH_4}{1\,mole\,H_2O} = 5.25\,moles\,CH_4\,(calc\ and\ corr)$$

b) $CaO + 2\,HCl \rightarrow CaCl_2 + H_2O$

$$5.25\,moles\,H_2O \times \frac{1\,mole\,CaCl_2}{1\,mole\,H_2O} = 5.25\,moles\,CaCl_2\,(calc\ and\ corr)$$

c) $4 NH_3 + 3 O_2 \rightarrow 2 N_2 + 6 H_2O$

$$5.25 \text{ moles } H_2O \times \frac{2 \text{ moles } N_2}{6 \text{ moles } H_2O} = 1.75 \text{ moles } N_2 \text{ (calc and corr)}$$

d) $2 HCl + Ba(OH)_2 \rightarrow BaCl_2 + 2 H_2O$

$$5.25 \text{ moles } H_2O \times \frac{1 \text{ mole } BaCl_2}{2 \text{ moles } H_2O} = 2.625 \text{ (calc)} = 2.62 \text{ moles } BaCl_2 \text{ (corr)}$$

10.71 a) $1.00 \text{ mole } H_2O \times \dfrac{4 \text{ moles } HNO_3}{2 \text{ moles } H_2O} \times \dfrac{63.02 \text{ g } HNO_3}{1 \text{ mole } HNO_3} = 126.04 \text{ (calc)} = 126 \text{ g } HNO_3 \text{ (corr)}$

b) $1.00 \text{ mole } H_2O \times \dfrac{6 \text{ moles } HNO_3}{3 \text{ moles } H_2O} \times \dfrac{63.02 \text{ g } HNO_3}{1 \text{ mole } HNO_3} = 126.04 \text{ (calc)} = 126 \text{ g } HNO_3 \text{ (corr)}$

c) $1.00 \text{ mole } H_2O \times \dfrac{1 \text{ mole } HNO_3}{2 \text{ moles } H_2O} \times \dfrac{63.02 \text{ g } HNO_3}{1 \text{ mole } HNO_3} = 31.51 \text{ (calc)} = 31.5 \text{ g } HNO_3 \text{ (corr)}$

d) $1.00 \text{ mole } H_2O \times \dfrac{10 \text{ moles } HNO_3}{3 \text{ moles } H_2O} \times \dfrac{63.02 \text{ g } HNO_3}{1 \text{ mole } HNO_3} = 210.06666 \text{ (calc)} = 210. \text{ g } HNO_3 \text{ (corr)}$

10.73 a) $1.772 \text{ g } CaO \times \dfrac{1 \text{ mole } CaO}{56.08 \text{ g } CaO} \times \dfrac{2 \text{ moles } HNO_3}{1 \text{ mole } CaO} \times \dfrac{63.02 \text{ g } HNO_3}{1 \text{ mole } HNO_3} = 3.98257632 \text{ (calc)}$

$$= 3.983 \text{ g } HNO_3 \text{ (corr)}$$

b) $1.772 \text{ g } PCl_5 \times \dfrac{1 \text{ mole } PCl_5}{208.22 \text{ g } PCl_5} \times \dfrac{4 \text{ moles } H_2O}{1 \text{ mole } PCl_5} \times \dfrac{18.02 \text{ g } H_2O}{1 \text{ mole } H_2O} = 0.6134173 \text{ (calc)}$

$$= 0.6134 \text{ g } H_2O \text{ (corr)}$$

c) $1.772 \text{ g } CuCl_2 \times \dfrac{1 \text{ mole } CuCl_2}{134.45 \text{ g } CuCl_2} \times \dfrac{2 \text{ moles } NaOH}{1 \text{ mole } CuCl_2} \times \dfrac{40.00 \text{ g } NaOH}{1 \text{ mole } NaOH} = 1.05436965 \text{ (calc)}$

$$= 1.054 \text{ g } NaOH \text{ (corr)}$$

d) $1.772 \text{ g } Na_2S \times \dfrac{1 \text{ mole } Na_2S}{78.04 \text{ g } Na_2S} \times \dfrac{2 \text{ moles } AgC_2H_3O_2}{1 \text{ mole } Na_2S} \times \dfrac{166.92 \text{ g } AgC_2H_3O_2}{1 \text{ mole } AgC_2H_3O_2}$

$$= 7.580272681 \text{ (calc)} = 7.580 \text{ g } AgC_2H_3O_2 \text{ (corr)}$$

10.75 a) $1.50 \text{ moles } C \times \dfrac{1 \text{ mole } SiO_2}{3 \text{ moles } C} \times \dfrac{60.09 \text{ g } SiO_2}{1 \text{ mole } SiO_2} = 30.045 \text{ (calc)} = 30.0 \text{ g } SiO_2 \text{ (corr)}$

b) $1.37 \text{ moles } SiO_2 \times \dfrac{2 \text{ moles } CO}{1 \text{ mole } SiO_2} \times \dfrac{28.01 \text{ g } CO}{1 \text{ mole } CO} = 76.7474 \text{ (calc)} = 76.7 \text{ g } CO \text{ (corr)}$

c) $3.33 \text{ moles } CO \times \dfrac{1 \text{ mole } SiC}{2 \text{ moles } CO} \times \dfrac{40.10 \text{ g } SiC}{1 \text{ mole } SiC} = 66.7665 \text{ (calc)} = 66.8 \text{ g } SiC \text{ (corr)}$

d) $0.575 \text{ mole } SiC \times \dfrac{3 \text{ moles } C}{1 \text{ mole } SiC} \times \dfrac{12.01 \text{ g } C}{1 \text{ mole } C} = 20.71725 \text{ (calc)} = 20.7 \text{ g } C \text{ (corr)}$

10.77 a) $4.50 \text{ moles } CO_2 \times \dfrac{2 \text{ moles } LiOH}{1 \text{ mole } CO_2} \times \dfrac{23.95 \text{ g } LiOH}{1 \text{ mole } LiOH} = 215.55 \text{ (calc)} = 216 \text{ g } LiOH \text{ (corr)}$

b) $3.00 \times 10^{24} \text{ molecules } CO_2 \times \dfrac{1 \text{ mole } CO_2}{6.022 \times 10^{23} \text{ molecules } CO_2} \times \dfrac{2 \text{ moles } LiOH}{1 \text{ mole } CO_2} \times \dfrac{23.95 \text{ g } LiOH}{1 \text{ mole } LiOH}$

$$= 238.62504 \text{ (calc)} = 239 \text{ g } LiOH \text{ (corr)}$$

c) $10.0 \text{ g } H_2O \times \dfrac{1 \text{ mole } H_2O}{18.02 \text{ g } H_2O} \times \dfrac{2 \text{ moles LiOH}}{1 \text{ mole } H_2O} \times \dfrac{23.95 \text{ g LiOH}}{1 \text{ mole LiOH}} = 26.581576 \text{ (calc)}$

$$= 26.6 \text{ g LiOH (corr)}$$

d) $10.0 \text{ g } Li_2CO_3 \times \dfrac{1 \text{ mole } Li_2CO_3}{73.89 \text{ g } Li_2CO_3} \times \dfrac{2 \text{ moles LiOH}}{1 \text{ mole } Li_2CO_3} \times \dfrac{23.95 \text{ g LiOH}}{1 \text{ mole LiOH}} = 6.4826093 \text{ (calc)}$

$$= 6.48 \text{ g LiOH (corr)}$$

10.79 a) $25.00 \text{ g NaF} \times \dfrac{1 \text{ mole NaF}}{41.99 \text{ g NaF}} \times \dfrac{1 \text{ mole } Na_2SiO_3}{2 \text{ moles NaF}}$

$$= 0.29768992 \text{ (calc)} = 0.2977 \text{ mole } Na_2SiO_3 \text{ (corr)}$$

b) $27.00 \text{ g } H_2O \times \dfrac{1 \text{ mole } H_2O}{18.02 \text{ g } H_2O} \times \dfrac{8 \text{ moles HF}}{3 \text{ moles } H_2O} \times \dfrac{20.01 \text{ g HF}}{1 \text{ mole HF}}$

$$= 79.951165 \text{ (calc)} = 79.95 \text{ g HF (corr)}$$

c) $2.000 \text{ g } Na_2SiO_3 \times \dfrac{1 \text{ mole } Na_2SiO_3}{122.07 \text{ g } Na_2SiO_3} \times \dfrac{1 \text{ mole } H_2SiF_6}{1 \text{ mole } Na_2SiO_3} \times \dfrac{6.022 \times 10^{23} \text{ molecules } H_2SiF_6}{1 \text{ mole } H_2SiF_6}$

$$= 9.8667401 \times 10^{21} \text{ (calc)} = 9.867 \times 10^{21} \text{ molecules } H_2SiF_6 \text{ (corr)}$$

d) $50.00 \text{ g } Na_2SiO_3 \times \dfrac{1 \text{ mole } Na_2SiO_3}{122.07 \text{ g } Na_2SiO_3} \times \dfrac{8 \text{ moles HF}}{1 \text{ mole } Na_2SiO_3} \times \dfrac{20.01 \text{ g HF}}{1 \text{ mole HF}} = 65.568936 \text{ (calc)}$

$$= 65.57 \text{ g HF (corr)}$$

10.81 Balanced equation: $4 \text{ Al} + 3 \text{ O}_2 \rightarrow 2 \text{ Al}_2O_3$

$23.7 \text{ g } O_2 \times \dfrac{1 \text{ mole } O_2}{32.00 \text{ g } O_2} \times \dfrac{4 \text{ moles Al}}{3 \text{ moles } O_2} \times \dfrac{26.98 \text{ g Al}}{1 \text{ mole Al}} = 26.64275 \text{ (calc)} = 26.6 \text{ g Al (corr)}$

LIMITING REACTANT CALCULATIONS (SEC. 10.10)

10.83 $216 \text{ nuts} \times \dfrac{1 \text{ combination}}{3 \text{ nuts}} = 72 \text{ combinations} \qquad 284 \text{ bolts} \times \dfrac{1 \text{ combination}}{4 \text{ bolts}} = 71 \text{ combinations}$
The limiting reactant is the 284 bolts.

10.85 a) $2.00 \text{ moles Be} \times \dfrac{3 \text{ moles } H_2}{3 \text{ moles Be}} = 2.00 \text{ moles } H_2 \text{ (calc and corr)}$

$0.500 \text{ mole } NH_3 \times \dfrac{3 \text{ moles } H_2}{2 \text{ moles } NH_3} = 0.75 \text{ (calc)} = 0.750 \text{ mole } H_2 \text{ (corr)}$
The 0.500 mole NH_3 is the limiting reactant.

b) $3.00 \text{ moles Be} \times \dfrac{3 \text{ moles } H_2}{3 \text{ moles Be}} = 3.00 \text{ moles } H_2 \text{ (calc and corr)}$

$3.00 \text{ moles } NH_3 \times \dfrac{3 \text{ moles } H_2}{2 \text{ moles } NH_3} = 4.5 \text{ (calc)} = 4.50 \text{ moles } H_2 \text{ (corr)}$
The 3.00 moles of Be is the limiting reactant.

c) $3.00 \text{ g Be} \times \dfrac{1 \text{ mole Be}}{9.01 \text{ g Be}} \times \dfrac{3 \text{ moles } H_2}{3 \text{ moles Be}} = 0.33296337 \text{ (calc)} = 0.333 \text{ mole } H_2 \text{ (corr)}$

$0.100 \text{ mole } NH_3 \times \dfrac{3 \text{ moles } H_2}{2 \text{ moles } NH_3} = 0.15 \text{ (calc)} = 0.150 \text{ mole } H_2 \text{ (corr)}$
The 0.100 mole of NH_3 is the limiting reactant.

d) $20.00 \text{ g Be} \times \dfrac{1 \text{ mole Be}}{9.01 \text{ g Be}} \times \dfrac{3 \text{ moles H}_2}{3 \text{ moles Be}} = 2.219755827 \text{ (calc)} = 2.22 \text{ moles H}_2 \text{ (corr)}$

$60.00 \text{ g NH}_3 \times \dfrac{1 \text{ mole NH}_3}{17.04 \text{ g NH}_3} \times \dfrac{3 \text{ moles H}_2}{2 \text{ moles NH}_3} = 5.281690141 \text{ (calc)} = 5.282 \text{ moles H}_2 \text{ (corr)}$

The 20.00 g of Be is the limiting reactant.

10.87 Determine which reactant is present in excess.

a) $2 \text{ NO} + \text{O}_2 \rightarrow 2 \text{ NO}_2$

$10.00 \text{ g NO} \times \dfrac{1 \text{ mole NO}}{30.01 \text{ g NO}} \times \dfrac{2 \text{ moles NO}_2}{2 \text{ moles NO}} = 0.333222259 \text{ (calc)} = 0.3332 \text{ mole NO}_2 \text{ (corr)}$

$15.00 \text{ g O}_2 \times \dfrac{1 \text{ mole O}_2}{32.00 \text{ g O}_2} \times \dfrac{2 \text{ moles NO}_2}{1 \text{ mole O}_2} = 0.9375 \text{ mole NO}_2 \text{ (calc and corr)}$

O_2 is present in excess.

b) $3 \text{ H}_2 + \text{N}_2 \rightarrow 2 \text{ NH}_3$

$10.00 \text{ g H}_2 \times \dfrac{1 \text{ mole H}_2}{2.02 \text{ g H}_2} \times \dfrac{2 \text{ moles NH}_3}{3 \text{ moles H}_2} = 3.300330033 \text{ (calc)} = 3.30 \text{ moles NH}_3 \text{ (corr)}$

$15.00 \text{ g N}_2 \times \dfrac{1 \text{ mole N}_2}{28.02 \text{ g N}_2} \times \dfrac{2 \text{ moles NH}_3}{1 \text{ mole N}_2} = 1.070663812 \text{ (calc)} = 1.071 \text{ moles NH}_3 \text{ (corr)}$

H_2 is present in excess.

c) $2 \text{ KOH} + \text{CuSO}_4 \rightarrow \text{Cu(OH)}_2 + \text{K}_2\text{SO}_4$

$10.00 \text{ g KOH} \times \dfrac{1 \text{ mole KOH}}{56.11 \text{ g KOH}} \times \dfrac{1 \text{ mole Cu(OH)}_2}{2 \text{ moles KOH}}$

$\qquad = 0.089110675 \text{ (calc)} = 0.08911 \text{ mole Cu(OH)}_2 \text{ (corr)}$

$15.00 \text{ g CuSO}_4 \times \dfrac{1 \text{ mole CuSO}_4}{159.61 \text{ g CuSO}_4} \times \dfrac{1 \text{ mole Cu(OH)}_2}{1 \text{ mole CuSO}_4}$

$\qquad = 0.093979073 \text{ (calc)} = 0.09398 \text{ mole Cu(OH)}_2 \text{ (corr)}$

CuSO_4 is present in excess.

d) $\text{Zn} + 2 \text{ HNO}_3 \rightarrow \text{Zn(NO}_3)_2 + \text{H}_2$

$10.00 \text{ g Zn} \times \dfrac{1 \text{ mole Zn}}{65.38 \text{ g Zn}} \times \dfrac{1 \text{ mole Zn(NO}_3)_2}{1 \text{ mole Zn}}$

$\qquad = 0.152951973 \text{ (calc)} = 0.1530 \text{ mole Zn(NO}_3)_2 \text{ (corr)}$

$15.00 \text{ g HNO}_3 \times \dfrac{1 \text{ mole HNO}_3}{63.02 \text{ g HNO}_3} \times \dfrac{1 \text{ mole Zn(NO}_3)_2}{2 \text{ moles HNO}_3}$

$\qquad = 0.119009838 \text{ (calc)} = 0.1190 \text{ mole Zn(NO}_3)_2 \text{ (corr)}$

Zn is present in excess.

10.89 Determine the limiting reactant in each. $\text{N}_2 + 3 \text{ H}_2 \rightarrow 2 \text{ NH}_3$

a) $3.0 \text{ g N}_2 \times \dfrac{1 \text{ mole N}_2}{28.02 \text{ g N}_2} \times \dfrac{2 \text{ moles NH}_3}{1 \text{ mole N}_2} \times \dfrac{17.04 \text{ g NH}_3}{1 \text{ mole NH}_3} = 3.6488223 \text{ (calc)} = 3.6 \text{ g NH}_3 \text{ (corr)}$

$5.0 \text{ g H}_2 \times \dfrac{1 \text{ mole H}_2}{2.02 \text{ g H}_2} \times \dfrac{2 \text{ moles NH}_3}{3 \text{ moles H}_2} \times \dfrac{17.04 \text{ g NH}_3}{1 \text{ mole NH}_3} = 28.11881188 \text{ (calc)} = 28 \text{ g NH}_3 \text{ (corr)}$

N_2 is the limiting reactant, and 3.6 g of NH_3 can be produced.

b) $30.0 \text{ g N}_2 \times \dfrac{1 \text{ mole N}_2}{28.02 \text{ g N}_2} \times \dfrac{2 \text{ moles NH}_3}{1 \text{ mole N}_2} \times \dfrac{17.04 \text{ g NH}_3}{1 \text{ mole NH}_3} = 36.488223 \text{ (calc)} = 36.5 \text{ g NH}_3 \text{ (corr)}$

$10.0 \text{ g H}_2 \times \dfrac{1 \text{ mole H}_2}{2.02 \text{ g H}_2} \times \dfrac{2 \text{ moles NH}_3}{3 \text{ moles H}_2} \times \dfrac{17.04 \text{ g NH}_3}{1 \text{ mole NH}_3} = 56.2376237 \text{ (calc)} = 56.2 \text{ g NH}_3 \text{ (corr)}$

N_2 is the limiting reactant, and 36.5 g of NH_3 can be produced.

c) $50.0 \text{ g N}_2 \times \dfrac{1 \text{ mole N}_2}{28.02 \text{ g N}_2} \times \dfrac{2 \text{ moles NH}_3}{1 \text{ mole N}_2} \times \dfrac{17.04 \text{ g NH}_3}{1 \text{ mole NH}_3} = 60.8137045 \text{ (calc)} = 60.8 \text{ g NH}_3 \text{ (corr)}$

$8.00 \text{ g H}_2 \times \dfrac{1 \text{ mole H}_2}{2.02 \text{ g H}_2} \times \dfrac{2 \text{ moles NH}_3}{3 \text{ moles H}_2} \times \dfrac{17.04 \text{ g NH}_3}{1 \text{ mole NH}_3} = 44.990099 \text{ (calc)} = 45.0 \text{ g NH}_3 \text{ (corr)}$

H_2 is the limiting reactant, and 45.0 g of NH_3 can be produced.

d) $56 \text{ g N}_2 \times \dfrac{1 \text{ mole N}_2}{28.02 \text{ g N}_2} \times \dfrac{2 \text{ moles NH}_3}{1 \text{ mole N}_2} \times \dfrac{17.04 \text{ g NH}_3}{1 \text{ mole NH}_3} = 68.111349 \text{ (calc)} = 68 \text{ g NH}_3 \text{ (corr)}$

$12 \text{ g H}_2 \times \dfrac{1 \text{ mole H}_2}{2.02 \text{ g H}_2} \times \dfrac{2 \text{ moles NH}_3}{3 \text{ moles H}_2} \times \dfrac{17.04 \text{ g NH}_3}{1 \text{ mole NH}_3} = 67.4851485 \text{ (calc)} = 67 \text{ g NH}_3 \text{ (corr)}$

H_2 is the limiting reactant, and 67 g of NH_3 can be produced.

10.91 Determine the amount of excess reactant that remains. $N_2 + 3 H_2 \rightarrow 2 NH_3$

a) H_2 is the excess reactant.

$3.0 \text{ g N}_2 \times \dfrac{1 \text{ mole N}_2}{28.02 \text{ g N}_2} \times \dfrac{3 \text{ moles H}_2}{1 \text{ mole N}_2} \times \dfrac{2.02 \text{ g H}_2}{1 \text{ mole H}_2}$
$= 0.648822269 \text{ (calc)} = 0.65 \text{ g H}_2 \text{ used (corr)}$

$5.0 \text{ g H}_2 \text{ present} - 0.65 \text{ g H}_2 \text{ used} = 4.35 \text{ (calc)} = 4.4 \text{ g H}_2 \text{ excess (corr)}$

b) H_2 is the excess reactant.

$30.0 \text{ g N}_2 \times \dfrac{1 \text{ mole N}_2}{28.02 \text{ g N}_2} \times \dfrac{3 \text{ moles H}_2}{1 \text{ mole N}_2} \times \dfrac{2.02 \text{ g H}_2}{1 \text{ mole H}_2}$
$= 6.488222698 \text{ (calc)} = 6.49 \text{ g H}_2 \text{ used (corr)}$

$10.0 \text{ g H}_2 \text{ present} - 6.49 \text{ g H}_2 \text{ used} = 3.51 \text{ (calc)} = 3.5 \text{ g H}_2 \text{ excess (corr)}$

c) N_2 is the excess reactant.

$8.00 \text{ g H}_2 \times \dfrac{1 \text{ mole H}_2}{2.02 \text{ g H}_2} \times \dfrac{1 \text{ mole N}_2}{3 \text{ moles H}_2} \times \dfrac{28.02 \text{ g N}_2}{1 \text{ mole N}_2}$
$= 36.99009901 \text{ (calc)} = 37.0 \text{ g N}_2 \text{ used (corr)}$

$50.0 \text{ g N}_2 \text{ present} - 37.0 \text{ g N}_2 \text{ used} = 13.0 \text{ g N}_2 \text{ excess (calc and corr)}$

d) N_2 is the excess reactant.

$12 \text{ g H}_2 \times \dfrac{1 \text{ mole H}_2}{2.02 \text{ g H}_2} \times \dfrac{1 \text{ mole N}_2}{3 \text{ moles H}_2} \times \dfrac{28.02 \text{ g N}_2}{1 \text{ mole N}_2}$
$= 55.48514851 \text{ (calc)} = 55 \text{ g N}_2 \text{ used (corr)}$

$56 \text{ g N}_2 \text{ present} - 55 \text{ g N}_2 \text{ used} = 1 \text{ g N}_2 \text{ excess (calc and corr)}$

10.93 a) Balanced reaction: $6 A_2 + 2 B_2 \rightarrow 4 BA_3$; we have $6 A_2$ and $4 B_2$.

$6 A_2 \times \dfrac{4 BA_3}{6 A_2} = 4 BA_3$ and $4 B_2 \times \dfrac{4 BA_3}{2 B_2} = 8 BA_3$; therefore A_2 is limiting.

b) Balanced reaction: $3 A_2 + 3 B_2 \rightarrow 6 AB$; we have $3 A_2$ and $6 B_2$.

$3 A_2 \times \dfrac{6 AB}{3 A_2} = 6 AB$ and $6 B_2 \times \dfrac{6 AB}{3 B_2} = 9 AB$; therefore A_2 is limiting.

10.95   The limiting reactant will all react. Use the limiting reactant to determine the mass of the other reactant that reacts, and then find the unreacted excess reactant by difference. $4 \, Fe_3O_4 + O_2 \rightarrow 6 \, Fe_2O_3$

$$70.0 \text{ g Fe}_3O_4 \times \frac{1 \text{ mole Fe}_3O_4}{231.55 \text{ g Fe}_3O_4} \times \frac{6 \text{ moles Fe}_2O_3}{4 \text{ moles Fe}_3O_4} = 0.453465774 \text{ (calc)} = 0.453 \text{ mole Fe}_2O_3 \text{ (corr)}$$

$$12.0 \text{ g O}_2 \times \frac{1 \text{ mole O}_2}{32.00 \text{ g O}_2} \times \frac{6 \text{ moles Fe}_2O_3}{1 \text{ mole O}_2} = 2.25 \text{ moles Fe}_2O_3 \text{ (calc and corr)}$$

The Fe_3O_4 is the limiting reactant. There will be none left upon completion.
Calculate the mass of O_2 reacted:

$$0.454 \text{ mole Fe}_2O_3 \times \frac{1 \text{ mole O}_2}{6 \text{ moles Fe}_2O_3} \times \frac{32.00 \text{ g O}_2}{1 \text{ mole O}_2} = 2.4213333 \text{ (calc)} = 2.42 \text{ g O}_2 \text{ reacted (corr)}$$

Unreacted: $Fe_3O_4 = 0$
$$O_2 = 12.0 \text{ g} - 2.42 \text{ g} = 9.58 \text{ (calc)} = 9.6 \text{ g O}_2 \text{ unreacted (corr)}$$

10.97 $$8.00 \text{ g SCl}_2 \times \frac{1 \text{ mole SCl}_2}{102.96 \text{ g SCl}_2} \times \frac{1 \text{ mole SF}_4}{3 \text{ moles SCl}_2} = 0.025900025 \text{ (calc)} = 0.0259 \text{ mole SF}_4 \text{ (corr)}$$

$$4.00 \text{ g NaF} \times \frac{1 \text{ mole NaF}}{41.99 \text{ g NaF}} \times \frac{1 \text{ mole SF}_4}{4 \text{ moles NaF}} = 0.023815194 \text{ (calc)} = 0.0238 \text{ mole SF}_4 \text{ (corr)}$$

NaF is the limiting reactant.

$$0.0238 \text{ mole SF}_4 \times \frac{108.06 \text{ g SF}_4}{1 \text{ mole SF}_4} = 2.571828 \text{ (calc)} = 2.57 \text{ g SF}_4 \text{ (corr)}$$

$$0.0238 \text{ mole SF}_4 \times \frac{1 \text{ mole S}_2Cl_2}{1 \text{ mole SF}_4} \times \frac{135.02 \text{ g S}_2Cl_2}{1 \text{ mole S}_2Cl_2} = 3.213476 \text{ (calc)} = 3.21 \text{ g S}_2Cl_2 \text{ (corr)}$$

$$0.0238 \text{ mole SF}_4 \times \frac{4 \text{ moles NaCl}}{1 \text{ mole SF}_4} \times \frac{58.44 \text{ g NaCl}}{1 \text{ mole NaCl}} = 5.563488 \text{ (calc)} = 5.56 \text{ g NaCl \text{ (corr)}$$

THEORETICAL YIELD AND PERCENT YIELD (SEC. 10.11)

10.99 $2 \, Mg + O_2 \rightarrow 2 \, MgO$
Given: 125.6 g MgO isolated out of possible 172.2 g MgO
a) Theoretical yield = 172.2 g MgO
b) Actual yield = 125.6 g MgO
c) Percent yield $= \dfrac{\text{actual yield}}{\text{theoretical yield}} \times 100 = \dfrac{125.6 \text{ g MgO}}{172.2 \text{ g MgO}} \times 100 = 72.93844367 \text{(calc)}$

$$= 72.94\% \text{ yield (corr)}$$

10.101   Diagram III

10.103   Theoretical yield of HCl:

$$2.13 \text{ g H}_2 \times \frac{1 \text{ mole H}_2}{2.02 \text{ g H}_2} \times \frac{2 \text{ moles HCl}}{1 \text{ mole H}_2} \times \frac{36.46 \text{ g HCl}}{1 \text{ mole HCl}} = 76.890891 \text{ (calc)} = 76.9 \text{ g HCl (corr)}$$

$$\% \text{ yield} = \frac{74.30 \text{ g HCl}}{76.9 \text{ g HCl}} \times 100 = 96.6189857 \text{ (calc)} = 96.6\% \text{ (corr)}$$

10.105 Theoretical yield of Al_2S_3:

$$55.0 \text{ g Al} \times \frac{1 \text{ mole Al}}{26.98 \text{ g Al}} \times \frac{1 \text{ mole Al}_2S_3}{2 \text{ moles Al}} \times \frac{150.14 \text{ g Al}_2S_3}{1 \text{ mole Al}_2S_3} = 153.0337287 \text{ (calc)}$$

$$= 153 \text{ g Al}_2S_3 \text{ (corr)}$$

Actual yield = theoretical yield × % yield

$$= 153 \text{ g Al}_2S_3 \text{ (theor)} \times \frac{85.6 \text{ g Al}_2S_3 \text{ (actual)}}{100 \text{ g Al}_2S_3 \text{ (theor)}} = 130.968 \text{ (calc)} = 131 \text{ g Al}_2S_3 \text{ (corr)}$$

10.107 Find the limiting reactant.

$$35.0 \text{ g CO} \times \frac{1 \text{ mole CO}}{28.01 \text{ g CO}} \times \frac{2 \text{ moles CO}_2}{2 \text{ moles CO}} = 1.2495537 \text{ (calc)} = 1.25 \text{ moles CO}_2 \text{ (calc and corr)}$$

$$35.0 \text{ g O}_2 \times \frac{1 \text{ mole O}_2}{32.00 \text{ g O}_2} \times \frac{2 \text{ moles CO}_2}{1 \text{ mole O}_2} = 2.1875 \text{ (calc)} = 2.19 \text{ moles CO}_2 \text{ (corr)}$$

CO is the limiting reactant.

$$\text{Theoretical yield} = 1.25 \text{ moles CO}_2 \times \frac{44.01 \text{ g CO}_2}{1 \text{ mole CO}_2} = 55.0125 \text{ (calc)} = 55.0 \text{ g CO}_2 \text{ (corr)}$$

$$\text{Actual yield} = 55.0 \text{ g CO}_2 \text{ (theor)} \times \frac{57.8 \text{ g CO}_2 \text{ (actual)}}{100 \text{ g CO}_2 \text{ (theor)}} = 31.79 \text{ (calc)} = 31.8 \text{ g CO}_2 \text{ (corr)}$$

10.109 3 sequential reactions, $A \rightarrow B \rightarrow C \rightarrow D$. The percent yield for this synthesis is the product of the percent yields of each step in the multi-step process.

$$100. \text{ g A} \times \frac{92 \text{ g B}}{100. \text{ g A}} \times \frac{87 \text{ g C}}{100. \text{ g B}} \times \frac{43 \text{ g D}}{100. \text{ g C}} = 34.4172 \text{ (calc)} = 34 \text{ g D (corr)}$$

$$\frac{34 \text{ g D}}{100. \text{ g A}} \times 100 = 34\% \text{ yield}$$

10.111 For the reactions $A \rightarrow B \rightarrow C$, the percent yield for the overall reaction is 75.0%; if the percent yield for the reaction of $B \rightarrow C$ is 90.0%, assign the percent yield for the reaction of $A \rightarrow B$ a value of X.

$$100.0 \text{ g A} \times \frac{X \text{ g B}}{100.0 \text{ g A}} \times \frac{90.0 \text{ g C}}{100.0 \text{ g B}} = 75.0 \text{ g C}$$

$$X \times 90.0 \text{ g C} = 75.0 \text{ g C}, \ X = (75.0 \text{ g C}/90.0 \text{ g C}) \times 100 = 83.33333333 \text{ (calc)} = 83.3\% \text{ (corr)}$$

SIMULTANEOUS CHEMICAL REACTIONS (SEC. 10.12)

10.113 $82.5 \text{ g mixture} \times \dfrac{60.0 \text{ g ZnS}}{100 \text{ g mixture}} \times \dfrac{1 \text{ mole ZnS}}{97.44 \text{ g ZnS}} \times \dfrac{2 \text{ moles SO}_2}{2 \text{ moles ZnS}} \times \dfrac{64.06 \text{ g SO}_2}{1 \text{ mole SO}_2}$

$$= 32.54279557 \text{ (calc)} = 32.5 \text{ g SO}_2 \text{ (corr)}$$

$82.5 \text{ g mixture} \times \dfrac{40.0 \text{ g CuS}}{100 \text{ g mixture}} \times \dfrac{1 \text{ mole CuS}}{95.61 \text{ g CuS}} \times \dfrac{2 \text{ moles SO}_2}{2 \text{ moles CuS}} \times \dfrac{64.06 \text{ g SO}_2}{1 \text{ mole SO}_2}$

$$= 22.1104487 \text{ (calc)} = 22.1 \text{ g SO}_2 \text{ (corr)}$$

Total $SO_2 = 32.5 \text{ g} + 22.1 \text{ g} = 54.6 \text{ g (calc and corr)}$

10.115 $75.0 \text{ g mixture} \times \dfrac{70.0 \text{ g CH}_4}{100 \text{ g mixture}} \times \dfrac{1 \text{ mole CH}_4}{16.05 \text{ g CH}_4} \times \dfrac{2 \text{ moles O}_2}{1 \text{ mole CH}_4} \times \dfrac{32.00 \text{ g O}_2}{1 \text{ mole O}_2}$

$$= 209.34579 \text{ (calc)} = 209 \text{ g O}_2 \text{ (corr)}$$

$75.0 \text{ g mixture} \times \dfrac{30.0 \text{ g C}_2\text{H}_6}{100 \text{ g mixture}} \times \dfrac{1 \text{ mole C}_2\text{H}_6}{30.08 \text{ g C}_2\text{H}_6} \times \dfrac{7 \text{ moles O}_2}{2 \text{ moles C}_2\text{H}_6} \times \dfrac{32.00 \text{ g O}_2}{1 \text{ mole O}_2}$

$$= 83.776596 \text{ (calc)} = 83.8 \text{ g O}_2 \text{ (corr)}$$

$\text{Total} = 209 \text{ g O}_2 + 83.8 \text{ g O}_2 = 292.8 \text{ (calc)} = 293 \text{ g O}_2 \text{ (corr)}$

SEQUENTIAL CHEMICAL REACTIONS (SEC. 10.12)

10.117 a) $6.00 \text{ moles SO}_2 \times \dfrac{1 \text{ mole O}_2}{1 \text{ mole SO}_2} \times \dfrac{2 \text{ moles NaClO}_3}{3 \text{ moles O}_2} = 49 \text{ (calc)} = 4.00 \text{ moles NaClO}_3 \text{ (corr)}$

b) $20.0 \text{ g SO}_2 \times \dfrac{1 \text{ mole SO}_2}{64.06 \text{ g SO}_2} \times \dfrac{1 \text{ mole O}_2}{1 \text{ mole SO}_2} \times \dfrac{2 \text{ moles NaClO}_3}{3 \text{ moles O}_2} \times \dfrac{106.44 \text{ g NaClO}_3}{1 \text{ mole NaClO}_3}$

$$= 22.15423041 \text{ (calc)} = 22.2 \text{ g NaClO}_3 \text{ (corr)}$$

10.119 a) $2.00 \text{ moles N}_2 \times \dfrac{2 \text{ moles NO}}{1 \text{ mole N}_2} \times \dfrac{2 \text{ moles NO}_2}{2 \text{ moles NO}} \times \dfrac{4 \text{ moles HNO}_3}{4 \text{ moles NO}_2} = 4 \text{ (calc)}$

$$= 4.00 \text{ moles HNO}_3 \text{ (corr)}$$

b) $2.00 \text{ g N}_2 \times \dfrac{1 \text{ mole N}_2}{28.02 \text{ g N}_2} \times \dfrac{2 \text{ moles NO}}{1 \text{ mole N}_2} \times \dfrac{2 \text{ moles NO}_2}{2 \text{ moles NO}} \times \dfrac{4 \text{ moles HNO}_3}{4 \text{ moles NO}_2} \times \dfrac{63.02 \text{ g HNO}_3}{1 \text{ mole HNO}_3}$

$$= 8.9964311 \text{ (calc)} = 9.00 \text{ g HNO}_3 \text{ (corr)}$$

10.121 $5.00 \text{ g I}_2 \times \dfrac{1 \text{ mole I}_2}{253.80 \text{ g I}_2} \times \dfrac{2 \text{ moles FeI}_2}{2 \text{ moles I}_2} \times \dfrac{2 \text{ moles AgI}}{1 \text{ mole FeI}_2} \times \dfrac{1 \text{ mole AgNO}_3}{1 \text{ mole AgI}} \times \dfrac{169.88 \text{ g AgNO}_3}{1 \text{ mole AgNO}_3}$

$$= 6.6934594 \text{ (calc)} = 6.69 \text{ g AgNO}_3 \text{ (corr)}$$

10.123 $2 \text{ NaClO}_3 \rightarrow 2 \text{ NaCl} + 3 \text{ O}_2$

$\underline{+\ 3 \text{ S} + 3 \text{ O}_2 \rightarrow 3 \text{ SO}_2}$

$2 \text{ NaClO}_3 + 3 \text{ S} \rightarrow 2 \text{ NaCl} + 2 \text{ O}_2 + 3 \text{ SO}_2$

10.125 $2 \text{ N}_2 + 2 \text{ O}_2 \rightarrow 4 \text{ NO}$

$+\ 4 \text{ NO} + 2 \text{ O}_2 \rightarrow 4 \text{ NO}_2$

$\underline{4 \text{ NO}_2 + 2 \text{ H}_2\text{O} + \text{O}_2 \rightarrow 4 \text{ HNO}_3}$

$2\text{N}_2 + 2 \text{ NO}_2 + 5 \text{ O}_2 \rightarrow 4 \text{ HNO}_3$

Multi-Concept Problems

10.127 a) $\text{HCl} + \text{NaOH} \rightarrow \text{NaCl} + \text{H}_2\text{O}$, consistent with law of conservation of mass

b) $\text{HNO}_3 + \text{Ca(OH)}_2 \rightarrow \text{Ca(NO}_3)_2 + \text{H}_2\text{O}$, not consistent with law of conservation of mass, atoms of N, O and H are different in the reactants and products.

c) $\text{Na}_2\text{SiO}_3 + 6 \text{ HF} \rightarrow \text{H}_2\text{SiF}_6 + 2 \text{ NaF} + 3 \text{ H}_2\text{O}$, not consistent with law of conservation of mass, atoms of F and H are different in the reactants and products.

d) $2 \text{ KHCO}_3 \rightarrow \text{K}_2\text{O} + \text{H}_2\text{O} + 2 \text{ CO}_2$, consistent with law of conservation of mass

10.129 $75.0 \text{ g } (NH_4)_2Cr_2O_7 \times \dfrac{1 \text{ mole } (NH_4)_2Cr_2O_7}{252.10 \text{ g } (NH_4)_2Cr_2O_7} \times \dfrac{1 \text{ mole } N_2}{1 \text{ mole } (NH_4)_2Cr_2O_7} \times \dfrac{28.02 \text{ g } N_2}{1 \text{ mole } N_2}$

$= 8.3359778 \text{ (calc)} = 8.34 \text{ g } N_2 \text{ (corr)}$

$75.0 \text{ g } (NH_4)_2Cr_2O_7 \times \dfrac{1 \text{ mole } (NH_4)_2Cr_2O_7}{252.10 \text{ g } (NH_4)_2Cr_2O_7} \times \dfrac{4 \text{ moles } H_2O}{1 \text{ mole } (NH_4)_2Cr_2O_7} \times \dfrac{18.02 \text{ g } H_2O}{1 \text{ mole } H_2O}$

$= 21.443871 \text{ (calc)} = 21.4 \text{ g } H_2O \text{ (corr)}$

$75.0 \text{ g } (NH_4)_2Cr_2O_7 \times \dfrac{1 \text{ mole } (NH_4)_2Cr_2O_7}{252.10 \text{ g } (NH_4)_2Cr_2O_7} \times \dfrac{1 \text{ mole } Cr_2O_3}{1 \text{ mole } (NH_4)_2Cr_2O_7} \times \dfrac{152.00 \text{ g } Cr_2O_3}{1 \text{ mole } Cr_2O_3}$

$= 45.220151 \text{ (calc)} = 45.2 \text{ g } Cr_2O_3 \text{ (corr)}$

10.131 Obtain a conversion factor with grams of product and grams of H_2S by finding the mass of products that could be made from 1.000 g H_2S:

$1.000 \text{ g } H_2S \times \dfrac{1 \text{ mole } H_2S}{34.08 \text{ g } H_2S} \times \dfrac{2 \text{ moles } SO_2}{2 \text{ moles } H_2S} \times \dfrac{64.06 \text{ g } SO_2}{1 \text{ mole } SO_2}$

$= 1.879694836 \text{ (calc)} = 1.879 \text{ g } SO_2 \text{ (corr) and}$

$1.000 \text{ g } H_2S \times \dfrac{1 \text{ mole } H_2S}{34.08 \text{ g } H_2S} \times \dfrac{2 \text{ moles } H_2O}{2 \text{ moles } H_2S} \times \dfrac{18.02 \text{ g } H_2O}{1 \text{ mole } H_2O}$

$= 0.528755868 \text{ (calc)} = 0.5288 \text{ g } H_2O \text{ (corr)}$

Total: $1.879 \text{ g} + 0.5288 = 2.4078 \text{ (calc)} = 2.408 \text{ g product (corr)}$

The conversion factor is $\dfrac{2.408 \text{ g product}}{1.000 \text{ g } H_2S}$

$100.0 \text{ g product} \times \dfrac{1.000 \text{ g } H_2S}{2.408 \text{ g product}} = 41.5282392 \text{ (calc)} = 41.53 \text{ g } H_2S \text{ (corr)}$

10.133 $2.00 \text{ moles } KNO_2 \times \dfrac{4 \text{ moles } NO}{3 \text{ moles } KNO_2} = 2.6666666 \text{ (calc)} = 2.67 \text{ moles } NO \text{ (corr)}$

$2.00 \text{ moles } KNO_3 \times \dfrac{4 \text{ moles } NO}{1 \text{ mole } KNO_3} = 8 \text{ (calc)} = 8.00 \text{ moles } NO \text{ (corr)}$

$2.00 \text{ moles } Cr_2O_3 \times \dfrac{4 \text{ moles } NO}{1 \text{ mole } Cr_2O_3} = 8 \text{ (calc)} = 8.00 \text{ moles } NO \text{ (corr)}; KNO_2 \text{ is the limiting reactant.}$

$2.67 \text{ moles } NO \times \dfrac{30.01 \text{ g } NO}{1 \text{ mole } NO} = 80.1267 \text{ (calc)} = 80.1 \text{ g } NO \text{ (corr)}$

10.135 $1.33 \text{ g Cu} \times \dfrac{1 \text{ mole Cu}}{63.55 \text{ g Cu}} \times \dfrac{1 \text{ mole } CuSO_4}{1 \text{ mole Cu}} \times \dfrac{159.60 \text{ g } CuSO_4}{1 \text{ mole } CuSO_4} = 3.340173092 \text{ (calc)}$

$= 3.34 \text{ g } CuSO_4 \text{ (corr) and } \% CuSO_4 = \dfrac{3.34 \text{ g}}{7.53 \text{ g}} \times 100 = 44.35591 \text{ (calc)} = 44.4\% \text{ (corr)}$

10.137 $113.4 \text{ g } I_2O_5 \times \dfrac{1 \text{ mole } I_2O_5}{333.80 \text{ g } I_2O_5} \times \dfrac{12 \text{ moles } IF_5}{6 \text{ moles } I_2O_5} = 0.67944877 \text{ (calc)} = 0.6794 \text{ mole } IF_5 \text{ (corr)}$

$132.2 \text{ g } BrF_3 \times \dfrac{1 \text{ mole } BrF_3}{136.90 \text{ g } BrF_3} \times \dfrac{12 \text{ moles } IF_5}{20 \text{ moles } BrF_3} = 0.57940102 \text{ (calc)} = 0.5794 \text{ mole } IF_5 \text{ (corr)}$

BrF_3 is the limiting reactant. Theoretical yield of IF_5:

$0.5794 \text{ mole } IF_5 \times \dfrac{221.9 \text{ g } IF_5}{1 \text{ mole } IF_5} = 128.56886 \text{ (calc)} = 128.6 \text{ g } IF_5 \text{ (corr)}$

$\% \text{ yield } IF_5 = \dfrac{97.0 \text{ g}}{128.6 \text{ g}} \times 100 = 75.4276835443 \text{ (calc)} = 75.4\% \text{ (corr)}$

10.139 Mass of $NaHCO_3$ present:

$0.873 \text{ g } H_2O \times \dfrac{1 \text{ mole } H_2O}{18.02 \text{ g } H_2O} \times \dfrac{2 \text{ moles } NaHCO_3}{1 \text{ mole } H_2O} \times \dfrac{84.01 \text{ g } NaHCO_3}{1 \text{ mole } NaHCO_3}$
$\qquad\qquad\qquad\qquad\qquad = 8.139925638 \text{ (calc)} = 8.14 \text{ g } NaHCO_3 \text{ (corr)}$

Mass of $CaCO_3$ present: $(13.20 - 8.14) \text{ g } CaCO_3 = 5.06 \text{ g } CaCO_3 \text{ (calc and corr)}$

$\% \ CaCO_3: \dfrac{5.06 \text{ g}}{13.20 \text{ g}} \times 100 = 38.3333 \text{ (calc)} = 38.3\% \text{ (corr)}$

Answers to Multiple-Choice Practice Test

MC 10.1 c	**MC 10.2** c	**MC 10.3** b	**MC 10.4** e	**MC 10.5** d	**MC 10.6** d
MC 10.7 b	**MC 10.8** e	**MC 10.9** e	**MC 10.10** d	**MC 10.11** a	**MC 10.12** b
MC 10.13 d	**MC 10.14** b	**MC 10.15** c	**MC 10.16** c	**MC 10.17** d	**MC 10.18** b
MC 10.19 d	**MC 10.20** c				

CHAPTER ELEVEN
States of Matter

PHYSICAL STATES OF MATTER (SECS. 11.1 AND 11.2)

11.1   a) True, both solids and liquids have definite volume.
b) True, thermal expansion for a liquid is generally greater than that of its corresponding solid.
c) False, the compressibility of a gas is generally greater than that of the corresponding liquid.
d) False, the density of a solid is greater than that of the corresponding gas.

11.3   a) both gases   b) liquid, solid   c) both gases   d) both solids

11.5   a) 3, H_2, N_2, Ne   b) 2, Br_2, Hg

KINETIC MOLECULAR THEORY OF MATTER (SECS. 11.3–11.6)

11.7   a) potential   b) kinetic   c) potential   d) potential

11.9   a) direct; the average velocity increases as the temperature increases and vice versa
b) potential (attractive)
c) direct; the higher the temperature, the higher the disruptive forces
d) all three; the disruptive forces are predominant in gases and the cohesive forces are more dominant in solids

11.11   a) solid   b) liquid   c) solid   d) solid or liquid

11.13   a) true   b) true   c) true   d) true

11.15   a) solid and liquid   b) solid   c) gas   d) gas

PHYSICAL CHANGES OF STATE (SEC. 11.8)

11.17   a) Evaporation is endothermic, freezing is exothermic, not the same thermicity.
b) Melting is endothermic, deposition is exothermic, not the same thermicity.
c) Freezing is exothermic, condensation is exothermic, same thermicity.
d) Sublimation is endothermic, evaporation is endothermic, same thermicity.

11.19   a) different, solid, liquid   b) different, liquid, solid   c) different, solid, liquid   d) same, gases

11.21   a) not opposite   b) not opposite   c) not opposite   d) not opposite

HEAT ENERGY AND SPECIFIC HEAT (SEC. 11.9)

11.23   a) $345 \text{ kJ} \times \dfrac{10^3 \text{ J}}{1 \text{ kJ}} = 3.45 \times 10^5 \text{ J}$ (calc and corr)

b) $345 \text{ kJ} \times \dfrac{1 \text{ kcal}}{4.184 \text{ kJ}} = 82.45697897$ (calc) $= 82.5 \text{ kcal}$ (corr)

c) $345 \text{ kJ} \times \dfrac{1 \text{ kcal}}{4.184 \text{ kJ}} \times \dfrac{10^3 \text{ cal}}{1 \text{ kcal}} = 82456.97897 \text{ (calc)} = 82,500 \text{ cal (corr)}$

d) $345 \text{ kJ} \times \dfrac{1 \text{ Cal}}{4.184 \text{ kJ}} = 82.45697897 \text{ (calc)} = 82.5 \text{ Cal (corr)}$

11.25   a) $2.0 \text{ calories} \times \dfrac{4.184 \text{ joules}}{1 \text{ calorie}} = 8.368 \text{ (calc)} = 8.4 \text{ joules (corr)} \quad \therefore 2.0 \text{ calories} > 2.0 \text{ joules}$

b) $1.0 \text{ kilocalorie} \times \dfrac{10^3 \text{ calories}}{1 \text{ kilocalorie}} = 1000 \text{ (calc)} = 1.0 \times 10^3 \text{ calories (corr)}$

$\therefore 1.0 \text{ kilocalorie} > 92 \text{ calories}$

c) $100 \text{ Calorie} \times \dfrac{10^3 \text{ calories}}{1 \text{ Calorie}} = 100,000 \text{ (calc)} = 1.0 \times 10^5 \text{ calories (corr)}$

$\therefore 100 \text{ Calories} > 100 \text{ calories}$

d) $2.3 \text{ Calories} \times \dfrac{10^3 \text{ calories}}{1 \text{ Calorie}} \times \dfrac{1 \text{ kilocalorie}}{10^3 \text{ calories}} = 2.3 \text{ kilocalories (calc and corr)}$

$\therefore 2.3 \text{ Calories} < 1000 \text{ kilocalories}$

11.27   Using values of specific heat from Table 11.2, $\Delta T = (35.0 - 25.0)^\circ\text{C} = 10.0^\circ\text{C}$

a) $\dfrac{0.382 \text{ J}}{\text{g} \, ^\circ\text{C}} \times 40.0 \text{ g} \times 10.0^\circ\text{C} = 152.8 \text{ (calc)} = 153 \text{ J (corr)}$

b) $\dfrac{0.444 \text{ J}}{\text{g} \, ^\circ\text{C}} \times 40.0 \text{ g} \times 10.0^\circ\text{C} = 177.6 \text{ (calc)} = 178 \text{ J (corr)}$

c) $\dfrac{1.0 \text{ J}}{\text{g} \, ^\circ\text{C}} \times 40.0 \text{ g} \times 10.0^\circ\text{C} = 400 \text{ (calc)} = 4.0 \times 10^2 \text{ J (corr)}$

d) $\dfrac{2.42 \text{ J}}{\text{g} \, ^\circ\text{C}} \times 40.0 \text{ g} \times 10.0^\circ\text{C} = 968 \text{ J (calc and corr)}$

11.29   Using values of specific heat from Table 11.2, 7.73 g of substance, above 43.2°C.

a) $\dfrac{145 \text{ J}}{0.908 \, \dfrac{\text{J}}{\text{g} \, ^\circ\text{C}} \times 7.73 \text{ g}} = 20.6586843 \text{ (calc)} = 20.7^\circ\text{C (corr) above } 43.2^\circ\text{C} = 63.9^\circ\text{C}$

b) $\dfrac{-145 \text{ J}}{0.908 \, \dfrac{\text{J}}{\text{g} \, ^\circ\text{C}} \times 7.73 \text{ g}} = -20.6586843 \text{ (calc)} = 20.7^\circ\text{C (corr) below } 43.2^\circ\text{C} = 22.5^\circ\text{C}$

c) $\dfrac{325 \text{ J}}{0.908 \, \dfrac{\text{J}}{\text{g} \, ^\circ\text{C}} \times 7.73 \text{ g}} = 46.30394766 \text{ (calc)} = 46.3^\circ\text{C (corr) above } 43.2^\circ\text{C} = 89.5^\circ\text{C}$

d) $\dfrac{-525 \text{ J}}{0.908 \, \dfrac{\text{J}}{\text{g} \, ^\circ\text{C}} \times 7.73 \text{ g}} = -74.79868468 \text{ (calc)} = 74.8^\circ\text{C (corr) below } 43.2^\circ\text{C} = -31.6^\circ\text{C}$

11.31   Using values of specific heat from Table 11.2, $\Delta T = (95 - 20.0)°C = 75°C$, absorbs 371 J of energy.

a) $g = \dfrac{371 \text{ J}}{0.444 \dfrac{\text{J}}{\text{g °C}} \times 75°C} = 11.1411411 \text{ (calc)} = 11 \text{ g Fe (corr)}$

b) $g = \dfrac{371 \text{ J}}{0.388 \dfrac{\text{J}}{\text{g °C}} \times 75°C} = 12.74914089 \text{ (calc)} = 13 \text{ g Zn (corr)}$

c) $g = \dfrac{371 \text{ J}}{4.18 \dfrac{\text{J}}{\text{g °C}} \times 75°C} = 1.1834131078 \text{ (calc)} = 1.2 \text{ g water (corr)}$

d) $g = \dfrac{371 \text{ J}}{2.42 \dfrac{\text{J}}{\text{g °C}} \times 75°C} = 2.044077135 \text{ (calc)} = 2.0 \text{ g ethyl alcohol (corr)}$

11.33   Use heat (released) = mass × specific heat × change in temperature
a) cooling 10.0 g of water from 50°C to 20°C

$10.0 \text{ g} \times \dfrac{4.184 \text{ J}}{\text{g°C}} \times (20°C - 50°C) = -1255.2 \text{ (calc)} = -1300 \text{ J (corr)}$

cooling 15.0 g of water from 75°C to 25°C

$15.0 \text{ g} \times \dfrac{4.184 \text{ J}}{\text{g °C}} \times (25°C - 75°C) = -3138 \text{ (calc)} = -3100 \text{ J (corr)}$

The first change, cooling 10.0 g of water 30°C, gives off less heat energy than the second process, cooling 15.0 g of water 50°C.
b) cooling 10.0 g of water from 50°C to 20°C

$10.0 \text{ g} \times \dfrac{4.184 \text{ J}}{\text{g °C}} \times (20°C - 50°C) = -1255.2 \text{ (calc)} = -1300 \text{ J (corr)}$

cooling 20.0 g of copper from 37°C to 25°C

$20.0 \text{ g} \times \dfrac{0.382 \text{ J}}{\text{g °C}} \times (25°C - 37°C) = -91.68 \text{ (calc)} = -92 \text{ J (corr)}$

The first change, cooling 10.0 g of water 30°C, gives off more heat energy than the second process, cooling 20.0 g of copper 12°C.

11.35   The most commonly used units for specific heat are Joules per gram per degree Celsius $\left(\dfrac{\text{J}}{\text{g °C}} \right)$.

11.37   Given: Metals X and Z are at the same mass, with the same initial and final temperatures.

Specific heat of metal X $= \dfrac{0.36 \text{ J}}{\text{g °C}}$

Specific heat of metal Z $= \dfrac{0.34 \text{ J}}{\text{g °C}}$

The lower specific heat of metal Z leads to the prediction that it will take less energy and therefore less time to raise its temperature to the same final temperature as metal X.

11.39   a) $\dfrac{39.8\ \text{J}}{40.0\ \text{g} \times 3.0°\text{C}} = 0.3316666\ (\text{calc})\ = 0.33\ \dfrac{\text{J}}{\text{g}\,°\text{C}}\ (\text{corr})$

      b) $\dfrac{46.9\ \text{J}}{40.0\ \text{g} \times 3.0°\text{C}} = 0.3908333\ (\text{calc})\ = 0.39\ \dfrac{\text{J}}{\text{g}\,°\text{C}}\ (\text{corr})$

      c) $\dfrac{57.8\ \text{J}}{40.0\ \text{g} \times 3.0°\text{C}} = 0.4816666\ (\text{calc})\ = 0.48\ \dfrac{\text{J}}{\text{g}\,°\text{C}}\ (\text{corr})$

      d) $\dfrac{75.0\ \text{J}}{40.0\ \text{g} \times 3.0°\text{C}} = 0.625\ (\text{calc})\ = 0.62\ \dfrac{\text{J}}{\text{g}\,°\text{C}}\ (\text{corr})$

11.41   Heat capacity = mass × specific heat (copper)
      Rearranging the equation to isolate mass on one side gives:

$$\text{mass} = \dfrac{\text{heat capacity}}{\text{specific heat}}$$

$$\text{mass} = \dfrac{66.85\ \text{J}/°\text{C}}{0.382\ \text{J}/\text{g}\,°\text{C}} = 175\ \text{g copper (calc and corr)}$$

11.43   Heat capacity of 40.0 g gold = 40.0 g × 0.13 J/g °C = 5.2 J/°C (calc and corr)
      Heat capacity of 80.0 g copper = 80.0 g × 0.382 J/g °C = 30.56 (calc) = 30.6 J/°C (corr)
      ∴ heat capacity of 80.0 g copper is higher.

11.45   Specific heat $= \dfrac{28.7\ \dfrac{\text{J}}{°\text{C}}}{75.1\ \text{g}} = 0.3821571238\ (\text{calc})\ = 0.382\ \dfrac{\text{J}}{\text{g}\,°\text{C}}\ (\text{corr})$

ENERGY AND CHANGES OF STATE (SEC. 11.11)

11.47   a) heat of solidification   b) heat of condensation   c) heat of fusion   d) heat of vaporization

11.49   a) $50.0\ \text{g Al} \times \dfrac{393\ \text{J}}{\text{g Al}} = -19650\ (\text{calc})\ = -1.96 \times 10^4\ \text{J released (corr)}$

      b) $50.0\ \text{g steam} \times \dfrac{2260\ \text{J}}{\text{g steam}} = -113{,}000\ (\text{calc})\ \text{or} -1.13 \times 10^5\ \text{J released (corr)}$

      c) $50.0\ \text{g Cu} \times \dfrac{205\ \text{J}}{\text{g Cu}} = 10{,}250\ (\text{calc})\ = 1.02 \times 10^4\ \text{J absorbed (corr)}$

      d) $50.0\ \text{g H}_2\text{O} \times \dfrac{2260\ \text{J}}{\text{g H}_2\text{O}} = 113{,}000\ (\text{calc})\ \text{or}\ 1.13 \times 10^5\ \text{J absorbed (corr)}$

11.51   The same amount, −3075 J

11.53 $\dfrac{1.00\ \text{mole Na}_2\text{SO}_4 \times 80.93\ \dfrac{\text{kJ}}{\text{mole}}\ \text{Na}_2\text{SO}_4}{1.00\ \text{mole NaOH} \times 15.79\ \dfrac{\text{kJ}}{\text{mole}}\ \text{NaOH}} = 5.1253958\ (\text{calc})\ = 5.13\ \text{times as much (corr)}$

11.55 Heat absorbed (J) = heat of fusion (J/g) × mass (g)
Results for both substances:

1. Rearranging the equation to isolate heat of fusion on one side gives:

$$\text{Heat of fusion (J/g)} = \frac{\text{heat absorbed (J)}}{\text{mass (g)}}$$

$$\text{Heat of fusion (J/g)} = \frac{1273 \text{ J}}{6.21 \text{ g}} = 204.9919485 \text{ (calc)} = 205 \text{ J/g (corr)}$$

2. Rearranging the equation to isolate heat of fusion on one side gives:

$$\text{Heat of fusion (J/g)} = \frac{\text{heat absorbed (J)}}{\text{mass (g)}}$$

$$\text{Heat of fusion (J/g)} = \frac{914 \text{ J}}{4.46 \text{ g}} = 204.9327354 \text{ (calc)} = 205 \text{ J/g (corr)}$$

Since both substances have the same heat of fusion as copper, the identity of the substance is copper.

11.57 $7.00 \text{ g Na} \times \dfrac{1 \text{ mole Na}}{22.99 \text{ g Na}} \times \dfrac{2.40 \text{ kJ}}{1 \text{ mole Na}} \times \dfrac{10^3 \text{ J}}{1 \text{ kJ}} = 730.7525 \text{ (calc)} = 7.31 \times 10^2 \text{ J (corr)}$

HEAT ENERGY CALCULATIONS (SEC. 11.12)

11.59

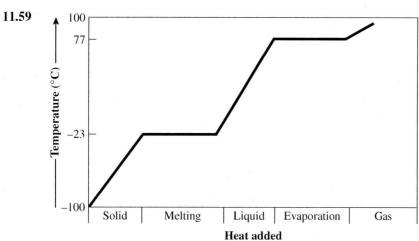

11.61

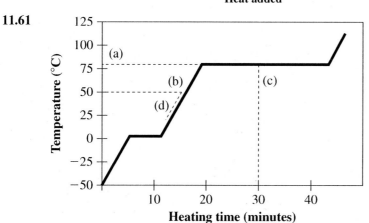

    a) approximate boiling point = 80°C

    b) at 50°C, substance is a liquid

    c) after 30 minutes, the physical state of the substance is part liquid and part gas

    d) as the substance is heated from 20°C to 50°C, the liquid warms

11.63   specific heat of ice = 2.09 J/g °C heat of fusion = 334 J/g specific heat of steam = 2.03 J/g °C

    specific heat of water = 4.18 J/g °C heat of vaporization = 2260 J/g

    a) Step 1: Heat ice from −18°C to −3°C.

           Q = 52.0 g ice × 2.09 J/g °C × 15°C = 1630.2 (calc) = 1.6×10^3 J (corr)

    b) Step 1: Heat ice from −18°C to 0°C.

           Q = 52.0 g × 2.09 J/g °C × 18°C = 1956.24 (calc) = 2.0×10^3 J (corr)

      Step 2: Melt ice at 0°C to water at 0°C.

           Q = 52.0 g × 334 J/g = 17368 (calc) = 17.4×10^3 J (corr)

      Step 3: Heat water from 0°C to 37°C.

           Q = 52.0 g × 4.18 J/g °C × 37°C = 8042.32 (calc) = 8.04×10^3 J (corr)

           Q total = Q ice + Q melting + Q water

                = 2.0×10^3 J + 17.4×10^3 J + 8.04×10^3 J = 27.44×10^3 (calc)

                = 27.4×10^3 J (corr)

    c) Step 1: Heat ice from −18°C to 0°C.

           Q = 52.0 g × 2.09 J/g °C × 18°C = 1956.24 (calc) = 2.0×10^3 J (corr)

      Step 2: Melt ice at 0°C to water at 0°C.

           Q = 52 g × 334 J/g = 17368 (calc) = 17.4×10^3 J (corr)

      Step 3: Heat water from 0°C to 100°C.

           52.0 g H_2O × 4.18 J/g °C × 100°C = 21,736 (calc) = 21.7×10^3 J (corr)

           Q total = Q ice + Q melting + Q water

                = 2.0×10^3 J + 17.4×10^3 J + 21.7×10^3 J = 41.1×10^3 J (calc and corr)

    d) Step 1: Heat ice from −18°C to 0°C.

           Q = 52.0 g × 2.09 J/g °C × 18°C = 1956.24 (calc) = 2.0×10^3 J (corr)

      Step 2: Melt ice at 0°C to water at 0°C.

           Q = 52 g × 334 J/g = 17368 (calc) = 17.4×10^3 J (corr)

      Step 3: Heat water from 0°C to 100°C.

           52.0 g H_2O × 4.18 J/g °C × 100°C = 21,736 (calc) = 21.7×10^3 J (corr)

      Step 4: Vaporize water at 100°C.

           52.0 g × 2260 J/g = 117,520 (calc) = 118×10^3 J (corr)

      Step 5: Heat steam from 100°C to 115°C.

           52.0 g × 2.03 J/g °C × 15.°C = 1583.4 (calc) = 1.6×10^3 J (corr)

           Q total = Q ice + Q melting + Q water + Q vap + Q steam

                = 2.0×10^3 J + 17.4×10^3 J + 21.7×10^3 J + 118×10^3 J + 1.6×10^3 J

                = 160.7×10^3 (calc) = 161×10^3 J (corr)

11.65   a) Step 1: Melt solid at 1530°C.

           35.2 g × 247 J/g = 8694.4 (calc) = 8.69×10^3 J (corr)

    b) Step 1: Heat solid at 25°C below melting point.

           35.2 g × 0.444 J/g °C × 25°C = 390.72 (calc) = 0.39×10^3 J (corr)

      Step 2: Melt solid at 1530°C.

           35.2 g × 247 J/g = 8694.4 (calc) = 8.69×10^3 J (corr)

           Q total = 0.39×10^3 J + 8.69×10^3 J = 9.08×10^3 J (calc and corr)

    c) Step 1: Heat solid from 1120°C to 1530°C.

           35.2 g × 0.444 J/g °C × 410.°C = 6407.808 (calc) = 6.41×10^3 J (corr)

      Step 2: Melt solid at 1530°C.

           35.2 g × 247 J/g = 8694.4 (calc) = 8.69×10^3 J (corr)

           Q total = 6.41×10^3 J + 8.69×10^3 J = 15.10×10^3 J (calc and corr)

d) Step 1: Heat solid from 27°C to 1530°C.

35.2 g × 0.444 J/g °C × 1503.°C = 23490.0864 (calc) = 23.4 × 10³ J (corr)

Step 2: Melt solid at 1530°C.

35.2 g × 247 J/g = 8694.4 (calc) = 8.69 × 10³ J (corr)

Q total = 23.4 × 10³ J + 8.69 × 10³ J = 32.09 × 10³ J (calc) = 32.1 × 10³ J (corr)

11.67 Step 1: solid (−135°C) → solid (−117°C)

15.0 g × 18°C × 0.97 J/g °C = 261.9 (calc) = 2.6 × 10² J (corr)

Step 2: solid (−117°C) → liquid (−117°C)

15.0 g × 109 J/g = 1635 J (calc) = 1.64 × 10³ J (corr)

Step 3: liquid (−117°C) → liquid (78°C)

15.0 g × 195°C × 2.3 J/g °C = 6727.5 (calc) = 6.7 × 10³ J (corr)

Step 4: liquid (78°C) → gas (78°C)

15.0 g × 837 J/g = 12,555 (calc) = 1.26 × 10⁴ J (corr)

Step 5: gas (78°C) → gas (95°C)

15.0 g × 17°C × 0.95 J/g °C = 242.25 J (calc) = 2.40 × 10² J (corr)

Total = 2.6 × 10² J + 1.64 × 10³ J + 6.7 × 10³ J + 1.26 × 10⁴ J + 2.40 × 10² J

= 21,440 (calc) = 2.14 × 10⁴ J (corr)

PROPERTIES OF LIQUIDS (SECS. 11.13–11.15)

11.69   a) boiling point   b) vapor pressure   c) boiling   d) boiling point

11.71   a) Increasing the temperature increases the average kinetic energy of the particles, enabling more molecules to evaporate.

b) The boiling point is lower; reactions occur more slowly at lower temperatures.

c) The boiling point is higher; reactions occur more rapidly at higher temperatures.

d) The particles leaving have higher than average kinetic energy. The particles remaining as liquid have a lower average kinetic energy and a lower temperature.

11.73   a) increase

b) no change—the rate of evaporation depends on the temperature, the cohesive forces present in the liquid, and the surface area where evaporation can occur. The concentration of air molecules above the liquid has a slight effect on the rate at which condensation of the vapor can occur. The *net* evaporation rate, the evaporation rate minus the condensation rate, may increase slightly at higher elevations, but the absolute rate of evaporation would not change.

c) increase—the liquid surface area has increased

d) no change

11.75   a) no change   b) decrease   c) no change   d) no change

11.77   a) increase

b) no change—the rate of evaporation depends on the temperature, the cohesive forces present in the liquid, and the surface area where evaporation can occur. The concentration of air molecules above the liquid has a slight effect on the rate at which condensation of the vapor can occur. The *net* evaporation rate, the evaporation rate minus the condensation rate, may increase slightly at higher elevations, but the absolute rate of evaporation would not change. This will make the vapor pressure increase very slightly.

c) no change

d) no change

11.79   B must have lower cohesive forces between particles than A to make B evaporate faster.

INTERMOLECULAR FORCES IN LIQUIDS (SEC. 11.16)

11.81   a) False, hydrogen bonds are an extra strong dipole–dipole interaction involving a hydrogen atom covalently bonded to a very small electronegative atom (F, O, or N) and an unshared pair of electrons on another very small electronegative atom (F, O, or N).

          b) True, dipole–dipole interactions are electrostatic in origin.

          c) True, all hydrogen bonds involve a hydrogen atom interacting with a lone pair of electrons on another atom.

          d) False, London forces are weak temporary dipole–dipole interactions between an atom or molecule and another atom or molecule.

11.83   The stronger the intermolecular forces, the higher the boiling point.

11.85   a) H_2: nonpolar, London forces

          b) HF: polar, hydrogen bonding, London forces

          c) CO: polar, dipole–dipole, London forces

          d) F_2: nonpolar, London forces

11.87   a) H_2: nonpolar, London forces

          b) HF: polar, hydrogen bonding is dominant

          c) CO: polar, dipole–dipole forces are dominant

          d) F_2: nonpolar, London forces

11.89   a) no hydrogen bonding   b) yes, hydrogen bonding

          c) yes, hydrogen bonding   d) no hydrogen bonding

11.91   Each water molecule, H_2O, can participate in *4 hydrogen bonds*. One hydrogen bond forms between each lone pair and the H atom on another H_2O molecule, and each H atom forms a hydrogen bond with a lone pair on another water molecule.

11.93   The strong electrostatic forces between ions in NiF_3 require much more energy to break than the weak intermolecular forces between polar molecules of NF_3, so at room temperature, NiF_3 is a solid while NF_3 is a gas.

11.95   a) F_2, larger molar mass   b) HF, hydrogen bonding

          c) CO, dipole–dipole   d) O_2, larger mass

11.97   a) As larger molecules have greater polarizability, $SiH_4 > CH_4$

          b) As larger molecules have greater polarizability, $SiCl_4 > SiH_4$

          c) As larger molecules have greater polarizability, $GeBr_4 > SiCl_4$

          d) As larger molecules have greater polarizability, $C_2H_4 > N_2$

HYDROGEN BONDING AND THE PROPERTIES OF WATER (SEC. 11.17)

11.99   Each water molecule forms hydrogen bonds to other water molecules. For water to boil, these hydrogen bonds must be broken, allowing the molecules to leave the liquid phase.

11.101   At 4°C, the distance between water molecules maximizes because of "balance" between random motion and hydrogen-bonding effects. Below this temperature, random motion decreases and hydrogen-bonding causes the molecules to move further apart, causing a density decrease.

11.103 The lower density of ice is due to the solid crystal lattice structure, which leaves open spaces between the molecules of H_2O.

11.105 Hydrogen-bonding causes water to have higher than normal heats of vaporization and condensation. Thus larger amounts of heat are absorbed during evaporation and released during condensation, which produces a temperature-moderating effect.

11.107 A measure of the inward force on the surface of a liquid caused by unbalanced intermolecular forces.

Multi-Concept Problems

11.109 a) $SnCl_4$ b) SnI_4 c) SnI_4 d) SnI_4

11.111 Let X = grams of ice added
Heat lost by the water is equal in magnitude but opposite in sign to the heat required to melt the ice.

Heat gained by ice to melt it: $X \text{ g} \times \dfrac{334 \text{ J}}{\text{g}} = 334 \cdot X \text{ J (calc and corr)}$

Heat lost by water: $40.0 \text{ g} \times \dfrac{4.18 \text{ J}}{\text{g} \, °\text{C}} \times -19.0°\text{C} = 3176.8 \text{ (calc)} = -3180 \text{ J (corr)}$

$-$Heat lost (water) $=$ heat gained (ice): $3180 \text{ J} = 334 \cdot X \text{ J}$ $X = 9.520958 \text{ (calc)} = 9.52 \text{ g (corr)}$

11.113 Let X = final temperature of water in °C
Heat lost by the water is equal in magnitude but opposite in sign to the heat gained by the ice.
Heat lost by water:
Water (80.0°C) $\rightarrow$ water (X°C)

$30.0 \text{ g} \times (80.0 - X)°\text{C} \times \dfrac{4.18 \text{ J}}{\text{g} \, °\text{C}} = (10{,}032 - 125.4 \cdot X) \text{ (calc)} = (1\overline{0{,}0}00 - 125 \cdot X) \text{ J (corr)}$

Heat gained by ice:
Step 1: ice $(-10°\text{C}) \rightarrow$ ice $(0°\text{C})$: $10.0 \text{ g} \times 10.0°\text{C} \times 2.09 \text{ J/g} \, °\text{C} = 209 \text{ J (calc and corr)}$
Step 2: ice $(0°\text{C}) \rightarrow$ water $(0°\text{C})$: $10.0 \text{ g} \times 334 \text{ J/g} = 3340 \text{ J (calc and corr)}$
Step 3: water $(0°\text{C}) \rightarrow$ water $(X°\text{C})$: $10.0 \text{ g} \times X°\text{C} \times 4.18 \text{ J/g} \, °\text{C} = 41.8 \cdot X \text{ J (calc and corr)}$
Total $= (209 \text{ J} + 3340 \text{ J} + 41.8 \cdot X \text{ J}) = ((3349 \text{ (calc) or } 3550 \text{ J (corr)}) + 41.8 \cdot X) \text{ J (calc and corr)}$
Heat lost (water) $=$ heat gained (ice): $(1\overline{0{,}0}00 - 125 \cdot X) \text{ J} = (3550 + 41.8 \cdot X) \text{ J}$
$6450 \text{ J} = (166.8 \cdot X) \text{ J (calc)}$ $6400 \text{ J} = (167 \cdot X) \text{ J (corr)}$ $X = 38.323353 \text{ (calc)} = 38.3°\text{C (corr)}$

11.115 Let X = specific heat of metal
Heat lost by the metal is equal in magnitude but opposite in sign to the heat gained by water.
Heat lost (metal) $=$ heat gained (water)

$500.0 \text{ g} \times X \dfrac{\text{J}}{\text{g} \, °\text{C}} \times 26.6°\text{C} = 100.0 \text{ g} \times \dfrac{4.18 \text{ J}}{\text{g} \, °\text{C}} \times 13.4°\text{C}$

$13{,}300 \cdot X = 5601.2 \text{ (calc)}$ $13{,}300 \cdot X = 56\overline{0}0 \text{ (corr)}$ $X = 0.42105263 \text{ (calc)} = 0.421 \text{ J/g} \, °\text{C (corr)}$

11.117 $4.3 \text{ min} \times \dfrac{60 \text{ sec}}{1 \text{ min}} \times \dfrac{136 \text{ J}}{\text{sec}} = 35{,}088 \text{ J (calc)} = 35{,}000 \text{ J (corr)}$

$\dfrac{35{,}000 \text{ J}}{25.3 \text{ g}} = 1383.3992 \text{ (calc)} = 1.4 \times 10^3 \text{ J/g (corr)}$

Answers to Multiple-Choice Practice Test

MC 11.1 a	**MC 11.2** c	**MC 11.3** e	**MC 11.4** e	**MC 11.5** a	**MC 11.6** e
MC 11.7 a	**MC 11.8** d	**MC 11.9** c	**MC 11.10** d	**MC 11.11** c	**MC 11.12** a
MC 11.13 d	**MC 11.14** c	**MC 11.15** b	**MC 11.16** e	**MC 11.17** d	**MC 11.18** a
MC 11.19 b	**MC 11.20** d				

CHAPTER TWELVE
Gas Laws

MEASUREMENT OF PRESSURE (SEC. 12.2)

12.1 a) $1.17 \text{ atmospheres (atm)} \times \dfrac{760 \text{ mm Hg}}{1 \text{ atm}} = 889.2 \text{ (calc)} = 890. \text{ mm Hg (corr)}$

b) $38.5 \text{ in. Hg} \times \dfrac{1 \text{ atm}}{29.92 \text{ in. Hg}} \times \dfrac{760 \text{ mm Hg}}{1 \text{ atm}} = 977.9411765 \text{ (calc)} = 978 \text{ mm Hg (corr)}$

c) $15.2 \text{ psi} \times \dfrac{1 \text{ atm}}{14.68 \text{ psi}} \times \dfrac{760 \text{ mm Hg}}{1 \text{ atm}} = 786.9209809 \text{ (calc)} = 787 \text{ mm Hg (corr)}$

d) $14.4 \text{ lb/in}^2 \text{ (psi)} \times \dfrac{1 \text{ atm}}{14.68 \text{ psi}} \times \dfrac{760 \text{ mm Hg}}{1 \text{ atm}} = 745.5040872 \text{ (calc)} = 746 \text{ mm Hg (corr)}$

12.3 a) $578 \text{ mm Hg} \times \dfrac{1 \text{ atm}}{760 \text{ mm Hg}} = 0.7605263158 \text{ (calc)} = 0.761 \text{ atm (corr)}, \text{ smaller than } 1.01 \text{ atm}$

b) $14.21 \text{ psi} \times \dfrac{1 \text{ atm}}{14.68 \text{ psi}} \times \dfrac{760 \text{ mm Hg}}{1 \text{ atm}} = 735.6675749 \text{ (calc)} = 735.7 \text{ mm Hg (corr)},$

smaller than 775 mm Hg

c) $29.92 \text{ in. Hg} \times \dfrac{1 \text{ atm}}{29.92 \text{ in. Hg}} \times \dfrac{760 \text{ mm Hg}}{1 \text{ atm}} = 760. \text{ mm Hg (calc and corr)}, \text{ equal to } 760 \text{ mm Hg}$

d) $3.57 \text{ atm} \times \dfrac{14.68 \text{ psi}}{1 \text{ atm}} = 52.4076 \text{ (calc)} = 52.4 \text{ psi (corr)}, \text{ greater than } 48.7 \text{ psi}$

12.5 $(762 + 237) \text{ mm Hg} = 999 \text{ mm Hg (calc and corr)}$

BOYLE'S LAW (SEC. 12.3)

12.7 a) increase b) increase c) decrease d) increase

12.9 For this problem, Boyle's Law is used to calculate the missing variable. $P_1V_1 = P_2V_2$

	P_1	V_1	P_2	V_2
	2.00 atm	2.00 L	4.00 atm	1.00 L
a)	2.67 atm	1.25 L	3.03 atm	1.10 L
b)	3.50 atm	1.73 L	2.75 atm	2.20 L
c)	1.75 atm	2.00 L	1.95 atm	1.79 L
d)	7.29 atm	3.25 L	3.00 atm	7.90 L

a) $P_2 = 2.67 \text{ atm} \times \dfrac{1.25 \text{ L}}{1.10 \text{ L}} = 3.034090909 \text{ (calc)} = 3.03 \text{ atm (corr)}$

b) $V_1 = \dfrac{2.75 \text{ atm}}{3.50 \text{ atm}} \times 2.20 \text{ L} = 1.728571429 \text{ (calc)} = 1.73 \text{ L (corr)}$

c) $V_2 = \dfrac{1.75 \text{ atm}}{1.95 \text{ atm}} \times 2.00 \text{ L} = 1.794871795 \text{ (calc)} = 1.79 \text{ L (corr)}$

d) $P_1 = 3.00 \text{ atm} \times \dfrac{7.90 \text{ L}}{3.25 \text{ L}} = 7.292307692 \text{ (calc)} = 7.29 \text{ atm (corr)}$

12.11   Boyle's law is used to determine the final volume that accompanies each pressure change.

$P_1 V_1 = P_2 V_2$, solving for $V_2 = \dfrac{P_1}{P_2} \times V_1$, $V_1 = 6.00 \text{ L}$, $V_{\text{change}} = V_2 - V_1$

a) pressure tripled: $V_2 = \dfrac{1\,P_1}{3\,P_2} \times 6.00 \text{ L} = 2.00 \text{ L (calc and corr)}$,

   volume reduced $= 2.00 \text{ L} - 6.00 \text{ L} = -4.00 \text{ L}$

b) pressure is reduced by a factor of 4: $V_2 = \dfrac{1\,P_1}{\frac{1}{4}P_2} \times 6.00 \text{ L} = 24.0 \text{ L (calc and corr)}$,

   volume increased $= 24.0 \text{ L} - 6.00 \text{ L} = 18.0 \text{ L}$

c) pressure is increased from 1.20 to 1.30 atm:

$$V_2 = \dfrac{1.20 \text{ atm}}{1.30 \text{ atm}} \times 6.00 \text{ L} = 5.538461538 \text{ (calc)} = 5.54 \text{ L (corr)}$$

   volume decreased $= 5.54 \text{ L} - 6.00 \text{ L} = -0.46 \text{ L}$

d) pressure is reduced from 1.20 to 1.10 atm:

$$V_2 = \dfrac{1.20 \text{ atm}}{1.10 \text{ atm}} \times 6.00 \text{ L} = 6.54545454 \text{ (calc)} = 6.55 \text{ L (corr)}$$

   volume is increased $= 6.55 \text{ L} - 6.00 \text{ L} = 0.55 \text{ L}$

12.13   Using Boyle's law, $P_1 V_1 = P_2 V_2$, if we double the pressure, the volume will decrease to one-half its original size. Diagram II represents the decrease in volume to one-half its original size with the same number of gas molecules.

CHARLES'S LAW (SEC. 12.4)

12.15   a) decrease   b) increase   c) decrease   d) increase

12.17   For this problem, Charles's Law is used to calculate the missing variable. $\dfrac{V_1}{T_1} = \dfrac{V_2}{T_2}$

	V_1	T_1	V_2	T_2
	2.00 L	27°C	4.00 L	327°C
a)	2.50 L	127°C	4.00 L	367°C
b)	3.30 L	165°C	4.05 L	265°C
c)	7.90 L	270°C	4.45 L	33°C
d)	0.974 L	100°C	1.00 L	110°C

a) $T_2 = 367°C + 273 = 640 K$

$$T_1 = \frac{2.50\ L}{4.00\ L} \times 640\ K = 400\ K\ (\text{calc and corr})$$

$$T_1 = 400\ K - 273 = 127°C$$

b) $T_1 = 165°C + 273 = 438\ K,\ T_2 = 265°C + 273 = 538\ K$

$$V_2 = 3.30\ L \times \frac{538\ K}{438\ K} = 4.053424658\ (\text{calc}) = 4.05\ L\ (\text{corr})$$

c) $T_1 = 270°C + 273 = 543\ K$

$$T_2 = \frac{4.45\ L}{7.90\ L} \times 543\ K = 305.8670886\ (\text{calc}) = 306\ K\ (\text{corr})$$

$$T_2 = 306\ K - 273 = 33°C$$

d) $T_1 = 100°C + 273 = 373\ K$

$$V_1 = 1.00\ L \times \frac{373\ K}{383\ K} = 0.973890339\ (\text{calc}) = 0.974\ L\ (\text{corr})$$

12.19 For this problem, Charles's Law is rearranged to calculate the new volume of a 6.00 L sample (V_1).

$$\frac{V_1}{T_1} = \frac{V_2}{T_2},\ \text{therefore}\ V_2 = V_1 \times \frac{T_2}{T_1},\ V_{\text{change}} = V_2 - V_1$$

a) Temperature is tripled: $V_2 = 6.00\ L \times \dfrac{3\ K}{1\ K} = 18.0\ L\ (\text{calc and corr})$

volume increased $= 18.0\ L - 6.00\ L = 12.0\ L$

b) Temperature is reduced by a factor of 4:

$$V_2 = 6.00\ L \times \frac{\frac{1}{4}\ K}{1\ K} = 1.50\ L\ (\text{calc and corr})$$

volume decreased $= 1.50\ L - 6.00\ L = -4.50\ L$

c) Temperature is increased: $T_1 = 120°C + 273 = 393\ K,$

$$T_2 = 130°C + 273 = 403\ K$$

$$V_2 = 6.00\ L \times \frac{403\ K}{393\ K} = 6.152671756\ (\text{calc}) = 6.15\ L\ (\text{corr})$$

volume increased $= 6.15\ L - 6.00\ L = 0.15\ L$

d) Temperature is decreased: $T_1 = 120°C + 273 = 393\ K,$

$$T_2 = 110°C + 273 = 383\ K$$

$$V_2 = 6.00\ L \times \frac{383\ K}{393\ K} = 5.847328244\ (\text{calc}) = 5.85\ L\ (\text{corr})$$

volume decreased $= 5.85\ L - 6.00\ L = -0.15\ L$

12.21 Using Charles's law, $V_1/T_1 = V_2/T_2$, decreasing the temperature while the pressure and moles of gas remain constant results in an equal decrease in the volume. Diagram II illustrates the decrease of volume by a factor of 2, while the number of moles of gas remains the same.

GAY-LUSSAC'S LAW (SEC. 12.5)

12.23 a) decrease   b) increase   c) decrease   d) increase

12.25 For this problem, Gay-Lussac's Law is rearranged to calculate the new variable. $\dfrac{P_1}{T_1} = \dfrac{P_2}{T_2}$

	P_1	T_1	P_2	T_2
	3.00 atm	127°C	6.00 atm	527°C
a)	1.11 atm	12°C	1.15 atm	22°C
b)	1.15 atm	24°C	0.929 atm	−33°C
c)	7.50 atm	627°C	5.50 atm	387°C
d)	2.13 atm	52°C	2.20 atm	62°C

a) $T_2 = 22°C + 273 = 295$ K

$$T_1 = \frac{1.11 \text{ atm}}{1.15 \text{ atm}} \times 295 \text{ K} = 284.7391304 \text{ (calc)} = 285 \text{ K (corr)}$$

$$T_1 = 285 \text{ K} - 273 = 12°C$$

b) $T_1 = 24°C + 273 = 297$ K, $T_2 = -33°C + 273 = 240$ K

$$P_2 = 1.15 \text{ atm} \times \frac{240 \text{ K}}{297 \text{ K}} = 0.929292929 \text{ (calc)} = 0.929 \text{ atm (corr)}$$

c) $T_1 = 627°C + 273 = 900$ K

$$T_2 = \frac{5.50 \text{ atm}}{7.50 \text{ atm}} \times 900 \text{ K} = 660 \text{ K (calc and corr)}$$

$$T_2 = 660 \text{ K} - 273 = 387°C$$

d) $T_1 = 52°C + 273 = 325$ K, $T_2 = 62°C + 273 = 335$ K

$$P_1 = 2.20 \text{ atm} \times \frac{325 \text{ K}}{335 \text{ K}} = 2.134328358 \text{ (calc)} = 2.13 \text{ atm (corr)}$$

12.27 For this problem, Gay Lussac's Law is rearranged to calculate the new pressure. $\dfrac{P_1}{T_1} = \dfrac{P_2}{T_2}$, therefore

$P_2 = P_1 \times \dfrac{T_2}{T_1}$ and $P_1 = 2.25$ atm, $P_{change} = P_2 - P_1$

a) Temperature is doubled: $P_2 = 2.25 \text{ atm} \times \dfrac{2 \text{ K}}{1 \text{ K}} = 4.50 \text{ atm (calc and corr)}$

pressure increased = 4.50 atm − 2.25 atm = 2.25 atm

b) Temperature is reduced by a factor of 2: $P_2 = 2.25 \text{ atm} \times \dfrac{\frac{1}{2} \text{ K}}{1 \text{ K}} = 1.125 \text{ (calc)} = 1.12 \text{ atm (corr)}$

pressure decreased = 1.12 atm − 2.25 atm = −1.13 atm

c) Temperature is increased: $T_1 = 125°C + 273 = 398$ K,

$$T_2 = 150°C + 273 = 423 \text{ K}$$

$$P_2 = 2.25 \times \frac{423 \text{ K}}{398 \text{ K}} = 2.391331658 \text{ (calc)} = 2.39 \text{ atm (corr)}$$

pressure increased = 2.39 atm − 2.25 atm = 0.14 atm

d) Temperature is decreased: $T_1 = 125°C + 273 = 398$ K,

$$T_2 = 100°C + 273 = 373 \text{ K}$$

$$P_2 = 2.25 \text{ atm} \times \frac{373 \text{ K}}{398 \text{ K}} = 2.108668342 \text{ (calc)} = 2.11 \text{ atm (corr)}$$

pressure decreased = 2.11 atm − 2.25 atm = −0.14 atm

12.29 $P_2 = P_1 \times \dfrac{T_2}{T_1}$ where $T_1 = 24°C + 273 = 297$ K and $T_2 = 485°C + 273 = 758$ K

$P_2 = 1.2 \text{ atm} \times \dfrac{758 \text{ K}}{297 \text{ K}} = 3.0626262 \text{ (calc)} = 3.1 \text{ atm (corr)}$

THE COMBINED GAS LAW (SEC. 12.6)

12.31 a) $\dfrac{P_1 V_1}{T_1} = \dfrac{P_2 V_2}{T_2}$ Multiply both sides by T_1 and T_2, and divide by P_1 and V_1.

$\dfrac{T_2 T_1}{P_1 V_1} \times \dfrac{P_1 V_1}{T_1} = \dfrac{P_2 V_2}{T_2} \times \dfrac{T_2 T_1}{P_1 V_1}$ gives $T_2 = T_1 \times \dfrac{P_2}{P_1} \times \dfrac{V_2}{V_1}$

b) $\dfrac{P_1 V_1}{T_1} = \dfrac{P_2 V_2}{T_2}$ Reverse the equation: $\dfrac{P_2 V_2}{T_2} = \dfrac{P_1 V_1}{T_1}$

Multiply both sides by T_2 and divide by P_1 and P_2.

$\dfrac{P_2 V_2}{T_2} \times \dfrac{T_2}{P_1 P_2} = \dfrac{P_1 V_1}{T_1} \times \dfrac{T_2}{P_1 P_2}$ gives $\dfrac{V_2}{P_1} = \dfrac{V_1}{P_2} \times \dfrac{T_2}{T_1}$

12.33 $V_2 = V_1 \times \dfrac{P_1}{P_2} \times \dfrac{T_2}{T_1}$ where $V_1 = 3.00$ mL; $P_1 = 0.980$ atm; $T_1 = 230°C + 273 = 503$ K

a) $P_2 = 2.00$ atm; $T_2 = 25°C + 273 = 298$ K

$V_2 = 3.00 \text{ mL} \times \dfrac{0.980 \text{ atm}}{2.00 \text{ atm}} \times \dfrac{298 \text{ K}}{503 \text{ K}} = 0.87089463 \text{ (calc)} = 0.871 \text{ mL (corr)}$

b) $P_2 = 3.00$ atm; $T_2 = -25°C + 273 = 248$ K

$V_2 = 3.00 \text{ mL} \times \dfrac{0.980 \text{ atm}}{3.00 \text{ atm}} \times \dfrac{248 \text{ K}}{503 \text{ K}} = 0.483180915 \text{ (calc)} = 0.483 \text{ mL (corr)}$

c) $P_2 = 5.00$ atm; $T_2 = -75°C + 273 = 198$ K

$V_2 = 3.00 \text{ mL} \times \dfrac{0.980 \text{ atm}}{5.00 \text{ atm}} \times \dfrac{198 \text{ K}}{503 \text{ K}} = 0.23145924 \text{ (calc)} = 0.231 \text{ mL (corr)}$

d) $P_2 = 7.75$ atm; $T_2 = -125°C + 273 = 148$ K

$V_2 = 3.00 \text{ mL} \times \dfrac{0.980 \text{ atm}}{7.75 \text{ atm}} \times \dfrac{148 \text{ K}}{503 \text{ K}} = 0.111619316 \text{ (calc)} = 0.112 \text{ mL (corr)}$

12.35 a) $V_2 = V_1 \times \dfrac{P_1}{P_2} \times \dfrac{T_2}{T_1} = 15.2 \text{ L} \times \dfrac{1.35 \text{ atm}}{3.50 \text{ atm}} \times \dfrac{308 \text{ K}}{306 \text{ K}} = 5.9011764 \text{ (calc)} = 5.90 \text{ L (corr)}$

b) $15.2 \text{ L} = 15,200 \text{ mL}$

$V_2 = V_1 \times \dfrac{P_1}{P_2} \times \dfrac{T_2}{T_1} = 15,200 \text{ mL} \times \dfrac{1.35 \text{ atm}}{6.70 \text{ atm}} \times \dfrac{370 \text{ K}}{306 \text{ K}} = 3703.2485 \text{ (calc)}$

$= 3.70 \times 10^3 \text{ mL (corr)}$

c) $P_2 = P_1 \times \dfrac{V_1}{V_2} \times \dfrac{T_2}{T_1} = 1.35 \text{ atm} \times \dfrac{15.2 \text{ L}}{10.0 \text{ L}} \times \dfrac{315 \text{ K}}{306 \text{ K}} = 2.1123529 \text{ (calc)} = 2.11 \text{ atm (corr)}$

d) $T_2 = T_1 \times \dfrac{P_2}{P_1} \times \dfrac{V_2}{V_1} = 306 \text{ K} \times \dfrac{7.00 \text{ atm}}{1.35 \text{ atm}} \times \dfrac{0.973 \text{ L}}{15.2 \text{ L}} = 101.56754 \text{ (calc)} = 102 \text{ K (corr)}$

Converting to Celsius: $102 \text{ K} - 273 = -171°C$

12.37 a) $375 \text{ mL} \times \dfrac{10^{-3} \text{ L}}{1 \text{ mL}} = 0.375 \text{ L}, P_2 = P_1 \times \dfrac{V_1}{V_2} = 1.03 \text{ atm} \times \dfrac{0.375 \text{ L}}{1.25 \text{ L}} = 0.309 \text{ atm (calc and corr)}$

$0.309 \text{ atm} \times \dfrac{760 \text{ mm Hg}}{1 \text{ atm}} = 234.84 \text{ (calc)} = 235 \text{ mm Hg (corr)}$

b) $T_1 = 27°C + 273 = 300. \text{ K}, \quad T_2 = T_1 \times \dfrac{V_2}{V_1} = 300: \text{K} \times \dfrac{1.25 \text{ L}}{0.375 \text{ L}} = 1\overline{0}00 \text{ K}$

Converting to Celsius: $1\overline{0}00 \text{ K} - 273 = 727°C \text{ (calc)} = 730°C \text{ (corr)}$

12.39 $T_2 = T_1 \times \dfrac{P_2}{P_1} \times \dfrac{V_2}{V_1}$ where $T_1 = 33°C + 273 = 306 \text{ K}.$

a) $P_2 = 3P_1; V_2 = 3V_1 \quad T_2 = 306 \text{ K} \times \dfrac{3P_1}{P_1} \times \dfrac{3V_1}{V_1} = 2754 \text{ (calc)} = 2750 \text{ K (corr)}$

Converting to Celsius: $2750 \text{ K} - 273 = 2477 \text{ (calc)} = 2480°C \text{ (corr)}$

b) $P_2 = \dfrac{1}{2} P_1 \text{ or } 2P_2 = P_1; V_2 = \dfrac{1}{2} V_1 \text{ or } 2V_2 = V_1$

$T_2 = 306 \text{ K} \times \dfrac{\frac{1}{2} P_1}{P_1} \times \dfrac{\frac{1}{2} V_1}{V_1} = 76.5 \text{ K (calc and corr)}$

Converting to Celsius: $76.5 \text{ K} - 273.2 = -196.7°C \text{ (calc and corr)}$

c) $P_2 = 3P_1, V_2 = \dfrac{1}{2} V_1 \quad T_2 = 306 \text{ K} \times \dfrac{3P_1}{P_1} \times \dfrac{\frac{1}{2} V_1}{V_1} = 459 \text{ K (calc and corr)}$

Converting to Celsius: $459 \text{ K} - 273 = 186°C \text{ (calc and corr)}$

d) $P_2 = \dfrac{1}{2} P_1, V_2 = 2V_1 \quad T_2 = 306 \text{ K} \times \dfrac{\frac{1}{2} P_1}{P_1} \times \dfrac{2V_1}{V_1} = 306 \text{ K (no change)}.$

Converting to Celsius: $306 \text{ K} - 273 = 33°C \text{ (calc and corr)}$

AVOGADRO'S LAW (SEC. 12.7)

12.41 $\dfrac{V_1}{n_1} = \dfrac{V_2}{n_2}$ Solving for $V_2 = n_2 \times \dfrac{V_1}{n_1}$

a) $2.00 \text{ moles} \times \dfrac{24.0 \text{ L}}{1.00 \text{ mole}} = 48 \text{ (calc)} = 48.0 \text{ L (corr)}$

b) $2.32 \text{ moles} \times \dfrac{24.0 \text{ L}}{1.00 \text{ mole}} = 55.68 \text{ L (calc)} = 55.7 \text{ L (corr)}$

c) $5.40 \text{ moles} \times \dfrac{24.0 \text{ L}}{1.00 \text{ mole}} = 129.6 \text{ (calc)} = 130. \text{ L (corr)}$

d) $0.500 \text{ mole} \times \dfrac{24.0 \text{ L}}{1.00 \text{ mole}} = 12 \text{ (calc)} = 12.0 \text{ L (corr)}$

12.43 For this problem, Avogadro's Law is rearranged to calculate the new variable. $\dfrac{V_1}{n_1} = \dfrac{V_2}{n_2}$

	V_1	n_1	V_2	n_2
	2.00 L	4.00 moles	4.00 L	8.00 moles
a)	2.67 L	3.33 moles	4.01 L	5.00 moles
b)	1.00 L	1.69 moles	3.21 L	5.41 moles
c)	2.50 L	3.50 moles	2.60 L	3.68 moles
d)	1.13 L	2.00 moles	5.50 L	9.75 moles

a) $V_2 = 2.67 \text{ L} \times \dfrac{5.00 \text{ moles}}{3.33 \text{ moles}} = 4.0090090 \text{ (calc)} = 4.01 \text{ L (corr)}$

b) $n_1 = \dfrac{1.00 \text{ L}}{3.21 \text{ L}} \times 5.41 \text{ moles} = 1.685358255 \text{ (calc)} = 1.69 \text{ moles (corr)}$

c) $n_2 = \dfrac{2.60 \text{ L}}{2.50 \text{ L}} \times 3.50 \text{ moles} = 3.64 \text{ moles (calc and corr)}$

d) $V_1 = 5.50 \text{ L} \times \dfrac{2.00 \text{ moles}}{9.75 \text{ moles}} = 1.128205128 \text{ (calc)} = 1.13 \text{ L (corr)}$

12.45 $\dfrac{V_1}{n_1} = \dfrac{V_2}{n_2}$ If we double the volume by increasing the moles of gas and keeping temperature and pressure constant, the moles of gas will be doubled. Diagram III represents a double volume and double the moles of gas.

12.47 Balloon, 7.83 moles of He, volume of 27.5 L at certain temperature and pressure.

a) $7.83 \text{ moles} + 1.00 \text{ mole} = 8.83 \text{ moles} \times \dfrac{27.5 \text{ L}}{7.83 \text{ moles}} = 31.01213282 \text{ (calc)} = 31.0 \text{ L (corr)}$

b) $7.83 \text{ moles} + 3.00 \text{ moles} = 10.83 \text{ moles} \times \dfrac{27.5 \text{ L}}{7.83 \text{ moles}} = 38.0363985 \text{ (calc)} = 38.0 \text{ L (corr)}$

c) $7.83 \text{ moles} - 1.00 \text{ mole} = 6.83 \text{ moles} \times \dfrac{27.5 \text{ L}}{7.83 \text{ moles}} = 23.98786718 \text{ (calc)} = 24.0 \text{ L (corr)}$

d) $7.83 \text{ moles} - 3.00 \text{ moles} = 4.83 \text{ moles} \times \dfrac{27.5 \text{ L}}{7.83 \text{ moles}} = 16.9636015 \text{ (calc)} = 17.0 \text{ L (corr)}$

12.49 1.00 mole Cl_2 gas, volume 24.0 L at certain temperature and pressure.

a) $2.00 \text{ g } Cl_2 \times \dfrac{1 \text{ mole } Cl_2}{70.90 \text{ g } Cl_2} \times \dfrac{24.0 \text{ L}}{1.00 \text{ mole } Cl_2} = 0.67700987306 \text{ (calc)} = 0.677 \text{ L (corr)}$

b) $2.32 \text{ g } Cl_2 \times \dfrac{1 \text{ mole } Cl_2}{70.90 \text{ g } Cl_2} \times \dfrac{24.0 \text{ L}}{1.00 \text{ mole } Cl_2} = 0.785331453 \text{ (calc)} = 0.785 \text{ L (corr)}$

c) $5.40 \text{ g } Cl_2 \times \dfrac{1 \text{ mole } Cl_2}{70.90 \text{ g } Cl_2} \times \dfrac{24.0 \text{ L}}{1.00 \text{ mole } Cl_2} = 1.827926658 \text{ (calc)} = 1.83 \text{ L (corr)}$

d) $0.500 \text{ g } Cl_2 \times \dfrac{1 \text{ mole } Cl_2}{70.90 \text{ g } Cl_2} \times \dfrac{24.0 \text{ L}}{1.00 \text{ mole } Cl_2} = 0.16925247 \text{ (calc)} = 0.169 \text{ L (corr)}$

12.51 Ideal gas law $PV = nRT$; if the pressure, volume, and temperature remain constant, the moles of gas will also remain constant.

a) 0.625 mole N_2 gas = 0.625 mole NO gas, therefore the volume will be 10.9 L

b) 0.625 mole N_2 gas = 0.625 mole N_2O gas, therefore the volume will be 10.9 L

c) 0.625 mole N_2 gas = 0.625 mole NO_2 gas, therefore the volume will be 10.9 L

d) 0.625 mole N_2 gas = 0.625 mole N_2O_5 gas, therefore the volume will be 10.9 L

12.53 $\dfrac{P_1 \times V_1}{n_1 \times T_1} = \dfrac{P_2 \times V_2}{n_2 \times T_2}$ $P_1 = 1.5$ atm, $V_1 = 2.0$ L, $n_1 = 0.500$ mole, $T_1 = 27°C + 273$ K $= 300$ K

$P_2 = 1.5$ atm $- 0.50$ atm $= 1.00$ atm

$V_2 = ?$

$n_2 = 0.500$ mole $- 0.100$ mole $= 0.400$ mole

$T_2 = 20°C + 27°C = 47°C$ $47°C + 273$ K $= 320$ K

$\dfrac{1.5 \text{ atm } 2.0 \text{ L}}{0.500 \text{ mole } 300 \text{ K}} = \dfrac{1.00 \text{ atm ? L}}{0.400 \text{ mole } 320 \text{ K}} = 2.56 \text{ (calc)} = 2.6 \text{ L (corr)}$

THE IDEAL GAS LAW (SEC. 12.9)

12.55 Solving $PV = nRT$ for V gives: $V = \dfrac{nRT}{P}$ where $n = 1.20$ moles

a) $P = 1$ atm, $T = 273$ K

$V = \dfrac{1.20 \text{ moles}}{1.00 \text{ atm}} \times 0.08206 \dfrac{\text{atm L}}{\text{mole K}} \times 273 \text{ K} = 26.882856 \text{ (calc)} = 26.9 \text{ L (corr)}$

b) $P = 1.54$ atm, $T = 73°C + 273 = 346$ K

$V = \dfrac{1.20 \text{ moles}}{1.54 \text{ atm}} \times 0.08206 \dfrac{\text{atm L}}{\text{mole K}} \times 346 \text{ K} = 22.124229 \text{ (calc)} = 22.1 \text{ L (corr)}$

c) $P = 15.0$ atm, $T = 525°C + 273 = 798$ K

$V = \dfrac{1.20 \text{ moles}}{15.0 \text{ atm}} \times 0.08206 \dfrac{\text{atm L}}{\text{mole K}} \times 798 \text{ K} = 5.2387104 \text{ (calc)} = 5.24 \text{ L (corr)}$

d) $P = 765$ mm Hg, $T = -23°C + 273 = 250$ K

$V = \dfrac{1.20 \text{ moles}}{765 \text{ mm Hg}} \times \dfrac{62.36 \text{ mm Hg} \cdot \text{L}}{\text{mole K}} \times 250 \text{ K} = 24.454902 \text{ (calc)} = 24.5 \text{ L (corr)}$

12.57 For this problem, the ideal gas law is rearranged to solve for each missing variable.

$PV = nRT, R = \dfrac{0.08206 \text{ atm L}}{\text{mole K}}$

	P	V	n	T
	10.9 atm	3.00 L	1.00 mole	127°C
a)	11.2 atm	2.00 L	1.00 mole	0°C
b)	4.00 atm	12.3 L	3.00 moles	−73°C
c)	4.75 atm	4.75 L	0.733 mole	102°C
d)	1.22 atm	3.73 L	0.100 mole	282°C

a) $T = 0°C + 273 = 273$ K $P = \dfrac{1.00 \text{ mole} \times \dfrac{0.08206 \text{ atm L}}{\text{mole K}} \times 273 \text{ K}}{2.00 \text{ L}}$

$= 11.20119 \text{ (calc)} = 11.2 \text{ atm (corr)}$

b) $T = -73°C + 273 = 200 \text{ K}$ $V = \dfrac{3.00 \text{ moles} \times \dfrac{0.08206 \text{ atm L}}{\text{mole K}} \times 200 \text{ K}}{4.00 \text{ atm}}$

$$= 12.309 \text{ (calc)} = 12.3 \text{ L (corr)}$$

c) $T = 102°C + 273 = 375 \text{ K}$ $n = \dfrac{4.75 \text{ atm} \times 4.75 \text{ L}}{375 \text{ K} \times \dfrac{0.08206 \text{ atm L}}{\text{mole K}}}$

$$= 0.733203347 \text{ (calc)} = 0.733 \text{ mole (corr)}$$

d) $T = \dfrac{1.22 \text{ atm} \times 3.73 \text{ L}}{0.100 \text{ mole} \times \dfrac{0.08206 \text{ atm L}}{\text{mole K}}} = 554.5454545 \text{ (calc)}$

$$= 555 \text{ K (corr)} \quad T = 555 \text{ K} - 273 = 282°C$$

12.59 $P = \dfrac{nRT}{V}$ where $V = 6.00 \text{ L}$, $T = 25°C + 273 = 298 \text{ K}$

a) $P = \dfrac{0.30 \text{ mole}}{6.00 \text{ L}} \times 0.08206 \dfrac{\text{atm L}}{\text{mole K}} \times 298 \text{ K} = 1.222694 \text{ (calc)} = 1.2 \text{ atm (corr)}$

b) $P = \dfrac{1.20 \text{ moles}}{6.00 \text{ L}} \times 0.08206 \dfrac{\text{atm L}}{\text{mole K}} \times 298 \text{ K} = 4.890776 \text{ (calc)} = 4.89 \text{ atm (corr)}$

c) $P = \dfrac{0.30 \text{ g N}_2}{6.00 \text{ L}} \times \dfrac{1 \text{ mole N}_2}{28.02 \text{ g N}_2} \times 0.08206 \dfrac{\text{atm L}}{\text{mole K}} \times 298 \text{ K}$

$$= 0.043636474 \text{ (calc)} = 0.044 \text{ atm (corr)}$$

d) $P = \dfrac{1.20 \text{ g N}_2}{6.00 \text{ L}} \times \dfrac{1 \text{ mole N}_2}{28.02 \text{ g N}_2} \times 0.08206 \dfrac{\text{atm L}}{\text{mole K}} \times 298 \text{ K}$

$$= 0.174545989 \text{ (calc)} = 0.175 \text{ atm (corr)}$$

12.61 $R = \dfrac{PV}{nT} = \dfrac{13.3 \text{ atm} \times 3.00 \text{ L}}{1.14 \text{ moles} \times 425 \text{ K}} = 0.082352941 \text{ (calc)} = 0.0824 \dfrac{\text{atm L}}{\text{mole K}} \text{ (corr)}$

12.63 $V = \dfrac{nRT}{P} = \dfrac{1.00 \text{ mole} \times 0.08206 \dfrac{\text{atm L}}{\text{mole K}} \times 400 \text{ K}}{0.908 \text{ atm}} = 36.14978 \text{ (calc)} = 36.1 \text{ L (corr)}$

12.65 $m = \dfrac{(M) PV}{RT}$, $T = 23°C + 273 = 296 \text{ K}$. For Cl_2, $(M) = 70.90 \text{ g/mole}$

$m = \dfrac{70.90 \text{ g}}{1 \text{ mole}} \times \dfrac{1.65 \text{ atm} \times 30.0 \text{ L}}{0.08206 \dfrac{\text{atm L}}{\text{mole K}} \times 296 \text{ K}} = 144.48681 \text{ (calc)} = 144 \text{ g Cl}_2 \text{ after adding (corr)}$

If there were 100.0 g Cl_2 initially, then $144 - 100.0 = 44 \text{ g Cl}_2$ added.

MODIFIED FORMS OF THE IDEAL GAS LAW EQUATION (SEC. 12.10)

12.67 $(M) = \dfrac{mRT}{PV}$, $P = 741 \text{ mm Hg} \times \dfrac{1 \text{ atm}}{760 \text{ mm Hg}} = 0.975 \text{ atm}$, $T = 33°C + 273 = 306 \text{ K}$

$\dfrac{1.305 \text{ g} \times 0.08206 \dfrac{\text{atm L}}{\text{mole K}} \times 306 \text{ K}}{1.20 \text{ L} \times 0.975 \text{ atm}} = 28.007709 \text{ (calc)} = 28.0 \text{ g/mole (corr)}$

12.69 $(M) = \dfrac{mRT}{PV}, T = 20°C + 273 = 293 \text{ K}$

$(M) = \dfrac{1.733 \text{ g} \times 0.08206 \dfrac{\text{atm L}}{\text{mole K}} \times 293 \text{ K}}{0.902 \text{ L} \times 1.65 \text{ atm}} = 27.996772387 \text{ (calc)} = 28.0 \text{ g/mole (corr)}$

CO has a molar mass of 28.01 g/mole; CO_2 has 44.01 g/mole. The gas is CO.

12.71 For a 10.0 g sample, the ideal gas law is rearranged to solve for each missing variable.

$PV = \dfrac{m}{M} RT, R = \dfrac{0.08206 \text{ atm L}}{\text{mole K}}$

	Molar mass (M)	P	V	T
	32.0 g/mole	3.85 atm	2.00 L	27°C
a)	28.0 g/mole	4.00 atm	2.00 L	0°C
b)	17.09 g/mole	1.50 atm	12.8 L	127°C
c)	114 g/mole	1.50 atm	1.50 L	40°C
d)	44.0 g/mole	4.00 atm	1.73 L	98°C

a) $T = 0°C + 273 = 273 \text{ K}$ $P = \dfrac{10.0 \text{ g} \times \dfrac{0.08206 \text{ atm L}}{\text{mole K}} \times 273 \text{ K}}{28.0 \text{ g/mole} \times 2.00 \text{ L}}$

$= 4.000425 \text{ (calc)} = 4.00 \text{ atm (corr)}$

b) $T = 127°C + 273 = 400 \text{ K}$ $V = \dfrac{10.0 \text{ g} \times \dfrac{0.08206 \text{ atm L}}{\text{mole K}} \times 400 \text{ K}}{17.09 \text{ g/mole} \times 1.50 \text{ atm}}$

$= 12.80436903 \text{ (calc)} = 12.8 \text{ L (corr)}$

c) $T = 40°C + 273 = 313 \text{ K}$ $M = \dfrac{10.0 \text{ g} \times \dfrac{0.08206 \text{ atm L}}{\text{mole K}} \times 313 \text{ K}}{1.50 \text{ L} \times 1.50 \text{ atm}}$

$= 114.1545778 \text{ (calc)} = 114 \text{ g/mole (corr)}$

d) $T = \dfrac{44.0 \text{ g/mole} \times 4.00 \text{ atm} \times 1.73 \text{ L}}{10.0 \text{ g} \times \dfrac{0.08206 \text{ atm L}}{\text{mole K}}} = 371.0455764 \text{ (calc)}$

$= 371 \text{ K (corr)} \; T = 371 \text{ K} - 273 = 98°C$

12.73 $d = \dfrac{m}{V} = \dfrac{(M) P}{RT}, (M \text{ H}_2\text{S}) = 34.08 \text{ g/mole}$

a) $T = 24°C + 273 = 297 \text{ K}.$

$d = \dfrac{34.08 \text{ g}}{1 \text{ mole}} \times \dfrac{675 \text{ mm Hg}}{62.36 \dfrac{\text{mm Hg L}}{\text{mole K}} \times 297 \text{ K}} = 1.24205493 \text{ (calc)} = 1.24 \text{ g/L (corr)}$

b) $T = 24°C + 273 = 297$ K.

$$d = \frac{34.08 \text{ g}}{1 \text{ mole}} \times \frac{1.20 \text{ atm}}{0.08206 \frac{\text{atm L}}{\text{mole K}} \times 297 \text{ K}} = 1.67800353 \text{ (calc)} = 1.68 \text{ g/L (corr)}$$

c) $T = 370°C + 273 = 643$ K.

$$d = \frac{34.08 \text{ g}}{1 \text{ mole}} \times \frac{1.30 \text{ atm}}{0.08206 \frac{\text{atm L}}{\text{mole K}} \times 643 \text{ K}} = 0.839654177 \text{ (calc)} = 0.840 \text{ g/L (corr)}$$

d) $T = -25°C + 273 = 248$ K.

$$d = \frac{34.08 \text{ g}}{1 \text{ mole}} \times \frac{452 \text{ mm Hg}}{62.36 \frac{\text{mm Hg} \cdot \text{L}}{\text{mole K}} \times 248 \text{ K}} = 0.996047921 \text{ (calc)} = 0.996 \text{ g/L (corr)}$$

12.75 $P = \dfrac{dRT}{(M)}$; $T = 47°C + 273 = 320$ K; $d = 1.00$ g/L

a) $(M) = \dfrac{28.02 \text{ g}}{1 \text{ mole}}$; $P = \dfrac{1.00 \text{ g}}{\text{L}} \times \dfrac{1 \text{ mole}}{28.02 \text{ g}} \times 0.08206 \dfrac{\text{atm L}}{\text{mole K}} \times 320 \text{ K}$

$$= 0.9371591 \text{ (calc)} = 0.937 \text{ atm (corr)}$$

b) $(M) = \dfrac{131.29 \text{ g}}{1 \text{ mole}}$; $P = \dfrac{1.00 \text{ g}}{\text{L}} \times \dfrac{1 \text{ mole}}{131.29 \text{ g}} \times 0.08206 \dfrac{\text{atm L}}{\text{mole K}} \times 320 \text{ K}$

$$= 0.2000091 \text{ (calc)} = 0.200 \text{ atm (corr)}$$

c) $(M) = \dfrac{54.45 \text{ g}}{1 \text{ mole}}$; $P = \dfrac{1.00 \text{ g}}{\text{L}} \times \dfrac{1 \text{ mole}}{54.45 \text{ g}} \times 0.08206 \dfrac{\text{atm L}}{\text{mole K}} \times 320 \text{ K}$

$$= 0.4822626 \text{ (calc)} = 0.482 \text{ atm (corr)}$$

d) $(M) = \dfrac{44.02 \text{ g}}{1 \text{ mole}}$; $P = \dfrac{1.00 \text{ g}}{\text{L}} \times \dfrac{1 \text{ mole}}{44.02 \text{ g}} \times 0.08206 \dfrac{\text{atm L}}{\text{mole K}} \times 320 \text{ K}$

$$= 0.5965288 \text{ (calc)} = 0.597 \text{ atm (corr)}$$

12.77 $T = 27°C + 273 = 300$ K; $(M) = \dfrac{dRT}{P}$

$$(M) = \frac{2.27 \text{ g}}{\text{L}} \times 0.08206 \frac{\text{atm L}}{\text{mole K}} \times \frac{300 \text{ K}}{2.00 \text{ atm}} = 27.94143 \text{ (calc)} = 27.9 \frac{\text{g}}{\text{mole}} \text{ (corr)}$$

Since CO has a molar mass of 28.01 g/mole and NO has 30.01 g/mole, the gas is CO.

12.79 $T = 25°C + 273 = 298$ K; $(M) = \dfrac{dRT}{P}$

$$(M) = \frac{1.35 \text{ g}}{\text{L}} \times 0.08206 \frac{\text{atm L}}{\text{mole K}} \times \frac{298 \text{ K}}{1.32 \text{ atm}} = 25.00965 \text{ (calc)} = 25.01 \frac{\text{g}}{\text{mole}} \text{ (corr)}$$

12.81 $T = \dfrac{P(M)}{dR} = 1.20 \text{ atm} \times \dfrac{30.01 \text{ g}}{1 \text{ mole}} \times \dfrac{1 \text{ L}}{1.25 \text{ g}} \times \dfrac{\text{mole K}}{0.08206 \text{ atm L}} = 351.0796978 \text{ (calc)}$

$$= 351 \text{ K (corr)} \quad °C = 351 \text{ K} - 273 = 78°C \text{ (calc and corr)}$$

VOLUMES OF GASES IN CHEMICAL REACTIONS (Sec. 12.11)

12.83 The whole-number combining ratio is $\dfrac{2.00 \text{ L O}_2}{1.60 \text{ L NH}_3} = \dfrac{5 \text{ L O}_2}{4 \text{ L NH}_3}$, which matches the coefficients in the

second reaction. $4 \text{ NH}_3(g) + 5 \text{ O}_2(g) \rightarrow 4 \text{ NO}(g) + 6 \text{ H}_2\text{O}(g)$

12.85 a) $1.30 \text{ L CO}_2 \times \dfrac{1 \text{ L C}_3\text{H}_4}{3 \text{ L CO}_2} = 0.43333333 \text{ (calc)} = 0.433 \text{ L C}_3\text{H}_8 \text{ (corr)}$

b) $1.30 \text{ L H}_2\text{O} \times \dfrac{1 \text{ L C}_3\text{H}_8}{4 \text{ L H}_2\text{O}} = 0.325 \text{ L C}_3\text{H}_8 \text{ (calc and corr)}$

12.87 0.75 total volume of products

At constant temperature and pressure, 1 L of C_3H_8 produces 3 L CO_2 and 4 L of H_2O, 7 L total products

$0.75 \text{ L total products} \times \dfrac{1 \text{ L C}_3\text{H}_8}{7 \text{ L products}} = 0.10714285 \text{ (calc)} = 0.11 \text{ L C}_3\text{H}_8 \text{ (corr)}$

VOLUMES OF GASES AND THE LIMITING REACTANT CONCEPT (SEC. 12.12)

12.89 Given the reaction $2 H_2S(g) + 3 O_2(g) \rightarrow 2 SO_2(g) + 2 H_2O(g)$, in which all gases are at the same temperature and pressure, determine the limiting reactant.

a) $1.25 \text{ L H}_2\text{S} \times \dfrac{2 \text{ L H}_2\text{O}}{2 \text{ L H}_2\text{S}} = 1.25 \text{ L H}_2\text{O (calc and corr)}$

$1.60 \text{ L O}_2 \times \dfrac{2 \text{ L H}_2\text{O}}{3 \text{ L O}_2} = 1.06666667 \text{ (calc)} = 1.07 \text{ L H}_2\text{O (corr)}$

$\qquad\qquad$ 1.60 L O_2 is the limiting reactant.

b) $2.50 \text{ L H}_2\text{S} \times \dfrac{2 \text{ L H}_2\text{O}}{2 \text{ L H}_2\text{S}} = 2.50 \text{ L H}_2\text{O (calc and corr)}$

$3.33 \text{ L O}_2 \times \dfrac{2 \text{ L H}_2\text{O}}{3 \text{ L O}_2} = 2.22 \text{ L H}_2\text{O (calc and corr)}$

$\qquad\qquad$ 3.33 L O_2 is the limiting reactant.

c) $5.00 \text{ L H}_2\text{S} \times \dfrac{2 \text{ L H}_2\text{O}}{2 \text{ L H}_2\text{S}} = 5.00 \text{ L H}_2\text{O (calc and corr)}$

$4.00 \text{ L O}_2 \times \dfrac{2 \text{ L H}_2\text{O}}{3 \text{ L O}_2} = 2.66666667 \text{ (calc)} = 2.67 \text{ L H}_2\text{O (corr)}$

$\qquad\qquad$ 4.00 L O_2 is the limiting reactant.

d) $3.00 \text{ L H}_2\text{S} \times \dfrac{2 \text{ L H}_2\text{O}}{2 \text{ L H}_2\text{S}} = 3.00 \text{ L H}_2\text{O (calc and corr)}$

$7.77 \text{ L O}_2 \times \dfrac{2 \text{ L H}_2\text{O}}{3 \text{ L O}_2} = 5.18 \text{ L H}_2\text{O (calc and corr)}$

$\qquad\qquad$ 3.00 L H_2S is the limiting reactant.

12.91 $CH_4 + 2 H_2O \rightarrow CO_2 + 4 H_2$

a) $45.0 \text{ L CH}_4 \times \dfrac{1 \text{ L CO}_2}{1 \text{ L CH}_4} = 45 \text{ (calc)} = 45.0 \text{ L CO}_2 \text{ (corr)}$

$45.0 \text{ L H}_2\text{O} \times \dfrac{1 \text{ L CO}_2}{2 \text{ L H}_2\text{O}} = 22.5 \text{ L CO}_2 \text{ (calc and corr)} \qquad \therefore 45.0 \text{ L H}_2\text{O is the limiting reactant.}$

$45.0 \text{ L H}_2\text{O} \times \dfrac{4 \text{ L H}_2}{2 \text{ L H}_2\text{O}} = 90 \text{ (calc)} = 90.0 \text{ L H}_2 \text{ (corr)}$

$\qquad\qquad \therefore$ 22.5 L CO_2 and 90.0 L H_2 are produced.

b) $45.0 \text{ L CH}_4 \times \dfrac{1 \text{ L CO}_2}{1 \text{ L CH}_4} = 45 \text{ (calc)} = 45.0 \text{ L CO}_2 \text{ (corr)}$

$66.0 \text{ L H}_2\text{O} \times \dfrac{1 \text{ L CO}_2}{2 \text{ L H}_2\text{O}} = 33 \text{ (calc)} = 33.0 \text{ L CO}_2 \text{ (corr)}$ $\therefore$ 66.0 L H$_2$O is the limiting reactant.

$66.0 \text{ L H}_2\text{O} \times \dfrac{4 \text{ L H}_2}{2 \text{ L H}_2\text{O}} = 132 \text{ L H}_2 \text{ (calc and corr)}$ $\therefore$ 33.0 L CO$_2$ and 132 L H$_2$ are produced.

c) $64.0 \text{ L CH}_4 \times \dfrac{1 \text{ L CO}_2}{1 \text{ L CH}_4} = (64) \text{ calc} = 64.0 \text{ L CO}_2 \text{ (corr)}$

$134 \text{ L H}_2\text{O} \times \dfrac{1 \text{ L CO}_2}{2 \text{ L H}_2\text{O}} = 67 \text{ (calc)} = 67.0 \text{ L CO}_2 \text{ (corr)}$ $\therefore$ 64.0 L CH$_4$ is the limiting reactant.

$64.0 \text{ L CH}_4 \times \dfrac{4 \text{ L H}_2}{1 \text{ L CH}_4} = 256 \text{ L H}_2 \text{ (calc and corr)}$ $\therefore$ 64.0 L CO$_2$ and 256 L H$_2$ are produced.

d) $16.0 \text{ L CH}_4 \times \dfrac{1 \text{ L CO}_2}{1 \text{ L CH}_4} = 16 \text{ (calc)} = 16.0 \text{ L CO}_2 \text{ (corr)}$

$32.0 \text{ L H}_2\text{O} \times \dfrac{1 \text{ L CO}_2}{2 \text{ L H}_2\text{O}} = 16 \text{ (calc)} = 16.0 \text{ L CO}_2 \text{ (corr)}$

$\therefore$ Both 16.0 L CH$_4$ and 32.0 L H$_2$O are limiting reactants.

$16.0 \text{ L CH}_4 \times \dfrac{4 \text{ L H}_2}{1 \text{ L CH}_4} = 64 \text{ (calc)} = 64.0 \text{ L H}_2 \text{ (corr)}$

or $32.0 \text{ L H}_2\text{O} \times \dfrac{4 \text{ L H}_2}{2 \text{ L H}_2\text{O}} = 64 \text{ (calc)} = 64.0 \text{ L H}_2 \text{ (corr)}$

$\therefore$ 16.0 L CO$_2$ and 64.0 L H$_2$ are produced.

MOLAR VOLUME OF A GAS (SEC. 12.13)

12.93 a) 1.20 atm, K = 33°C + 273 = 306 K

$V = \dfrac{nRT}{P} = 1 \text{ mole} \times \dfrac{0.08206 \text{ atm L}}{\text{mole K}} \times \dfrac{306 \text{ K}}{1.20 \text{ atm}} = 20.9253 \text{ (calc)} = 20.9 \text{ L (corr)}$

b) 1.20 atm, K = 45°C + 273 = 318 K

$V = \dfrac{nRT}{P} = 1 \text{ mole} \times \dfrac{0.08206 \text{ atm L}}{\text{mole K}} \times \dfrac{318 \text{ K}}{1.20 \text{ atm}} = 21.7459 \text{ (calc)} = 21.7 \text{ L (corr)}$

c) 1.00 atm, K = 123°C + 273 = 396 K

$V = \dfrac{nRT}{P} = 1 \text{ mole} \times \dfrac{0.08206 \text{ atm L}}{\text{mole K}} \times \dfrac{396 \text{ K}}{1.00 \text{ atm}} = 32.49576 \text{ (calc)} = 32.5 \text{ L (corr)}$

d) 2.00 atm, K = 123°C + 273 = 396 K

$V = \dfrac{nRT}{P} = 1 \text{ mole} \times \dfrac{0.08206 \text{ atm L}}{\text{mole K}} \times \dfrac{396 \text{ K}}{2.00 \text{ atm}} = 16.2478 \text{ (calc)} = 16.2 \text{ L (corr)}$

12.95   Molar volume $= V = \dfrac{nRT}{P}$, K $= 22°C + 273 = 295$ K, 2.00 atm

a) For N_2: $V = \dfrac{1 \text{ mole } \dfrac{0.08206 \text{ atm L}}{\text{mole K}} \, 295 \text{ K}}{2.00 \text{ atm}} = 12.10385 \text{ (calc)} = 12.1 \text{ L (corr)}$

b) For Ar: $V = \dfrac{1 \text{ mole } \dfrac{0.08206 \text{ atm L}}{\text{mole K}} \, 295 \text{ K}}{2.00 \text{ atm}} = 12.10385 \text{ (calc)} = 12.1 \text{ L (corr)}$

c) For SO_2: $V = \dfrac{1 \text{ mole } \dfrac{0.08206 \text{ atm L}}{\text{mole K}} \, 295 \text{ K}}{2.00 \text{ atm}} = 12.10385 \text{ (calc)} = 12.1 \text{ L (corr)}$

d) For SO_3: $V = \dfrac{1 \text{ mole } \dfrac{0.08206 \text{ atm L}}{\text{mole K}} \, 295 \text{ K}}{2.00 \text{ atm}} = 12.10385 \text{ (calc)} = 12.1 \text{ L (corr)}$

12.97   Molar volume of any gas at STP $= 22.41$ L; the molar volume does not depend on the identity of the gas, only the moles of gas.

For 1 mole of N_2, Ar, SO_2 and SO_3 gases:

$$V = \dfrac{1 \text{ mole } \dfrac{0.082057 \text{ atm L}}{\text{mole K}} \, 273.15 \text{ K}}{1.000 \text{ atm}} = 22.413869 \text{ (calc)} = 22.41 \text{ L (corr)}$$

12.99   Molar Volume is defined as the volume of 1 mole of gas.

	T	P	Molar Volume
	27°C	2.00 atm	12.3 L
a)	127°C	3.00 atm	10.9 L
b)	127°C	4.00 atm	8.21 L
c)	227°C	5.00 atm	8.21 L
d)	STP	STP	22.41 L

a) $T = 127°C + 273 = 400$ K

$$\text{Volume (1 mole)} = \dfrac{1 \text{ mole} \times \dfrac{0.08206 \text{ atm L}}{\text{mole K}} \times 400 \text{ K}}{3.00 \text{ atm}}$$

$$= 10.9413333 \text{ (calc)} = 10.9 \text{ L (corr)}$$

b) $T = 127°C + 273 = 400$ K

$$P = \dfrac{1 \text{ mole} \times \dfrac{0.08206 \text{ atm L}}{\text{mole K}} \times 400 \text{ K}}{8.21 \text{ L}}$$

$$= 3.998051157 \text{ (calc)} = 4.00 \text{ atm (corr)}$$

c) $T = \dfrac{5.00 \text{ atm} \times 8.21 \text{ L}}{1 \text{ mole} \times \dfrac{0.08206 \text{ atm L}}{\text{mole K}}} = 500.2437241 \text{ (calc)} = 500 \text{ K (corr)}$

$$T = 500 \text{ K} - 273 = 227°C$$

d) At STP, the volume of 1 mole of gas $= 22.41$ L.

12.101 Density $\dfrac{g}{L} = \dfrac{\dfrac{g}{mole}}{\dfrac{L}{mole}}$

a) Density $= \dfrac{\dfrac{17.04\ g}{1\ mole}}{\dfrac{17.21\ L}{1\ mole}} = 0.990122022\ (calc) = 0.9901\ \dfrac{g}{L}\ (corr)$

b) Density $= \dfrac{\dfrac{17.04\ g}{1\ mole}}{\dfrac{22.41\ L}{1\ mole}} = 0.760374832\ (calc) = 0.7604\ \dfrac{g}{L}\ (corr)$

c) Density $= \dfrac{\dfrac{17.04\ g}{1\ mole}}{\dfrac{23.34\ L}{1\ mole}} = 0.73007712\ (calc) = 0.7301\ \dfrac{g}{L}\ (corr)$

d) Density $= \dfrac{\dfrac{17.04\ g}{1\ mole}}{\dfrac{35.00\ L}{1\ mole}} = 0.486857142\ (calc) = 0.4869\ \dfrac{g}{L}\ (corr)$

12.103 a) $K = 25°C + 273 = 298\ K$, 1.20 atm

Molar Volume $= \dfrac{1\ mole\ \dfrac{0.08206\ atm\ L}{mole\ K}\ 298\ K}{1.20\ atm} = 20.3782333\ (calc) = 20.4\ \dfrac{L}{mole}\ (corr)$

Density $= \dfrac{\dfrac{44.01\ g\ CO_2}{1\ mole}}{\dfrac{20.4\ L}{1\ mole}} = 2.1573529\ (calc) = 2.16\ \dfrac{g}{L}\ (corr)$

b) $K = 35°C + 273 = 308\ K$, 1.33 atm

Molar Volume $= \dfrac{1\ mole\ \dfrac{0.08206\ atm\ L}{mole\ K}\ 308\ K}{1.33\ atm} = 19.0033684\ (calc) = 19.0\ \dfrac{L}{mole}\ (corr)$

Density $= \dfrac{\dfrac{28.01\ g\ CO}{1\ mole}}{\dfrac{19.0\ L}{1\ mole}} = 1.47421053\ (calc) = 1.47\ \dfrac{g}{L}\ (corr)$

c) At STP: Density $= \dfrac{\dfrac{16.05\ g\ CH_4}{1\ mole}}{\dfrac{22.41\ L}{1\ mole}} = 0.716198125$ (calc) $= 0.7162\ \dfrac{g}{L}$ (corr)

d) At STP: Density $= \dfrac{\dfrac{30.08\ g\ C_2H_6}{1\ mole}}{\dfrac{22.41\ L}{1\ mole}} = 1.342257921$ (calc) $= 1.342\ \dfrac{g}{L}$ (corr)

12.105 Since each gas will have the same molar volume at STP, density will be the greatest for the highest molecular weight.

a) O_3 b) PH_3 c) CO_2 d) F_2

12.107 Density $= \dfrac{Molar\ Mass\ (M)}{Molar\ Volume} = (M) = D \times Molar\ Volume$

a) $1.97\ \dfrac{g}{L} \times 22.41\ \dfrac{L}{mole} = 44.1477$ (calc) $= 44.1\ \dfrac{g}{mole}$ (corr)

b) $1.25\ \dfrac{g}{L} \times 22.41\ \dfrac{L}{mole} = 28.0125$ (calc) $= 28.0\ \dfrac{g}{mole}$ (corr)

c) $0.714\ \dfrac{g}{L} \times 22.41\ \dfrac{L}{mole} = 16.00074$ (calc) $= 16.0\ \dfrac{g}{mole}$ (corr)

d) $3.17\ \dfrac{g}{L} \times 22.41\ \dfrac{L}{mole} = 71.0397$ (calc) $= 71.0\ \dfrac{g}{mole}$ (corr)

12.109 Two flasks have identical volumes: one is filled with Ne, the other with He. Since the flasks have the same volume and are under the same temperature and pressure, the flask with Ne in it will have a larger mass. The molar mass of He (4.00 g/mole) is less than the molar mass of Ne (20.18 g/mole).

CHEMICAL CALCULATIONS USING MOLAR VOLUME (SEC. 12.14)

12.111 a) $23.7\ L \times \dfrac{1\ mole\ Ar}{22.41\ L} \times \dfrac{39.95\ g\ Ar}{1\ mole\ Ar} = 42.2496653$ (calc) $= 42.2\ g\ Ar$ (corr)

b) $23.7\ L \times \dfrac{1\ mole\ N_2O}{22.41\ L} \times \dfrac{44.02\ g\ N_2O}{1\ mole\ N_2O} = 46.55394913$ (calc) $= 46.6\ g\ N_2O$ (corr)

c) $23.7\ L \times \dfrac{1\ mole\ SO_3}{22.41\ L} \times \dfrac{80.06\ g\ SO_3}{1\ mole\ SO_3} = 84.66854083$ (calc) $= 84.7\ g\ SO_3$ (corr)

d) $23.7\ L \times \dfrac{1\ mole\ PH_3}{22.41\ L} \times \dfrac{34.00\ g\ PH_3}{1\ mole\ PH_3} = 35.95716198$ (calc) $= 36.0\ g\ PH_3$ (corr)

12.113 K $= 44°C + 273 = 317$ K, 1.15 atm, 23.7 L samples
Moles of gas present in each sample will be the same.

$n = \dfrac{1.15\ atm\ 23.7\ L}{\dfrac{0.08206\ atm\ L\ 317\ K}{mole\ K}} = 1.04774455$ (calc) $= 1.05$ moles gas (corr)

a) $1.05 \text{ moles} \times \dfrac{39.95 \text{ g Ar}}{1 \text{ mole Ar}} = 41.9475 \text{ (calc)} = 41.9 \text{ g Ar (corr)}$

b) $1.05 \text{ moles} \times \dfrac{44.02 \text{ g N}_2\text{O}}{1 \text{ mole N}_2\text{O}} = 46.221 \text{ (calc)} = 46.2 \text{ g N}_2\text{O (corr)}$

c) $1.05 \text{ moles} \times \dfrac{80.06 \text{ g SO}_3}{1 \text{ mole SO}_3} = 84.063 \text{ (calc)} = 84.1 \text{ g SO}_3 \text{ (corr)}$

d) $1.05 \text{ moles} \times \dfrac{17.04 \text{ g NH}_3}{1 \text{ mole NH}_3} = 17.892 \text{ (calc)} = 17.9 \text{ g NH}_3 \text{ (corr)}$

12.115 For a, b, c and d: $1.25 \text{ moles gas} \times \dfrac{22.41 \text{ L gas}}{1 \text{ mole gas}} = 28.0125 \text{ (calc)} = 28.0 \text{ L gas (corr)}$

Given 1.25 moles samples of each gas, at STP, 1 mole of any gas occupies 22.41 L. Therefore each gas will occupy 28.0 liters.

12.117 $46°C + 273 = 319 \text{ K}, 1.73 \text{ atm}$

$$V = \dfrac{1.25 \text{ moles} \dfrac{0.08206 \text{ atm L}}{\text{mole K}} 319 \text{ K}}{1.73 \text{ atm}} = 18.914118497 \text{ (calc)} = 18.9 \text{ L (corr)}$$

The volume will be the same for each sample of gas; it depends only on the temperature and pressure, not the identity of the gas. Each gas will therefore occupy 18.9 liters.

12.119 25.0 g of NO react with an excess of O_2 according to the following chemical equation:

$2 \text{ NO(g)} + O_2\text{(g)} \rightarrow 2 \text{ NO}_2\text{(g)}$

a) Liters of NO_2 produced at STP (NO is limiting):

$$25.0 \text{ g NO} \times \dfrac{1 \text{ mole NO}}{30.01 \text{ g NO}} \times \dfrac{2 \text{ moles NO}_2}{2 \text{ mole NO}} \times \dfrac{22.41 \text{ L NO}_2}{1 \text{ mole NO}_2}$$
$$= 18.66877707 \text{ (calc)} = 18.7 \text{ L NO}_2 \text{ (corr)}$$

b) Liters of NO_2 at 55°C and 2.50 atm produced;

$$25.0 \text{ g NO} \times \dfrac{1 \text{ mole NO}}{30.01 \text{ g NO}} \times \dfrac{2 \text{ moles NO}_2}{2 \text{ mole NO}} = 0.833055648 \text{ (calc)} = 0.833 \text{ moles NO}_2 \text{ (corr)}$$

$T = 55°C + 273 = 328 \text{ K}$

$$V = \dfrac{nRT}{P} = \dfrac{0.833 \text{ mole} \times \dfrac{0.08206 \text{ atm L}}{\text{mole K}} \times 328 \text{K}}{2.50 \text{ atm}} = 8.968304576 \text{ (calc)} = 8.97 \text{ L NO}_2 \text{ (corr)}$$

12.121 Method 1. Find moles of O_2. $\qquad\qquad T = 27°C + 273 = 300 \text{ K}$

$$n = \dfrac{1.00 \text{ atm} \times 25.0 \text{ L}}{0.08206 \dfrac{\text{atm L}}{\text{mole K}} \times 300 \text{ K}} = 1.0155171 \text{ (calc)} = 1.02 \text{ moles O}_2 \text{ (corr)}$$

$$1.02 \text{ moles O}_2 \times \dfrac{2 \text{ moles NO}}{1 \text{ mole O}_2} \times \dfrac{30.01 \text{ g NO}}{1 \text{ mole NO}} = 61.2204 \text{ (calc)} = 61.2 \text{ g NO (corr)}$$

Method 2. Find volume of O_2 at STP (P constant).

$$V_2 = \frac{V_1 T_2}{T_1} . \; V_2 = 25.0 \text{ L} \times \frac{273 \text{ K}}{300 \text{ K}} = 22.75 \text{ (calc)} = 22.8 \text{ L } O_2 \text{ at STP (corr)}$$

$$22.8 \text{ L } O_2 \times \frac{1 \text{ mole } O_2}{22.41 \text{ L } O_2} \times \frac{2 \text{ moles NO}}{1 \text{ mole } O_2} \times \frac{30.01 \text{ g NO}}{1 \text{ mole NO}} = 61.06453 \text{ (calc)} = 61.1 \text{ g NO (corr)}$$

Note: Rounding in intermediate steps causes the difference.

12.123 $\quad 12.0 \text{ g Mg} \times \dfrac{1 \text{ mole Mg}}{24.31 \text{ g Mg}} \times \dfrac{1 \text{ mole } H_2}{1 \text{ mole Mg}} = 0.49382716 \text{ (calc)} = 0.494 \text{ mole } H_2 \text{ (corr)}$

$$V = \frac{0.494 \text{ mole} \times 0.08206 \dfrac{\text{atm L}}{\text{mole K}} \times 296 \text{ K}}{0.980 \text{ atm}} = 12.244022 \text{ (calc)} = 12.2 \text{ L (corr)}$$

12.125 $\quad T = 450°C + 273 = 723 \text{ K.}$ $\qquad\qquad$ Conversion factor: $\dfrac{7 \text{ moles product gases}}{2 \text{ moles } NH_4NO_3}$

$$100.0 \text{ g } NH_4NO_3 \times \frac{1 \text{ mole } NH_4NO_3}{80.06 \text{ g } NH_4NO_3} \times \frac{7 \text{ moles product}}{2 \text{ moles } NH_4NO_3}$$
$$= 4.3717212 \text{ (calc)} = 4.37 \text{ moles product (corr)}$$

$$\text{Method 1: } V = \frac{nRT}{P} = \frac{4.37 \text{ moles}}{1 \text{ atm}} \times 0.08206 \frac{\text{atm L}}{\text{mole K}} \times 723 \text{ K} = 259.26939 \text{ (calc)} = 259 \text{ L (corr)}$$

$$\text{Method 2: } 4.37 \text{ moles} \times \frac{22.41 \text{ L}}{1 \text{ mole}} \times \frac{723 \text{ K}}{273 \text{ K}} = 259.35758 \text{ (calc)} = 259 \text{ L (corr)}$$

12.127 $\quad T = 38°C + 273 = 311 \text{ K} \quad n = \dfrac{645 \text{ mm Hg} \times 75.0 \text{ L}}{62.36 \dfrac{\text{mm Hg L}}{\text{mole K}} \times 311 \text{ K}} = 2.494333287 \text{ (calc)}$

$$= 2.49 \text{ moles NO (corr)}$$

$$2.49 \text{ moles NO} \times \frac{3 \text{ moles } NO_2}{1 \text{ mole NO}} = 7.47 \text{ moles } NO_2 \text{ (calc and corr)}$$

$$T = 21°C + 273 = 294 \text{ K} \quad V = \frac{7.47 \text{ moles} \times 0.08206 \dfrac{\text{atm L}}{\text{mole K}} \times 294 \text{ K}}{2.31 \text{ atm}} = 78.01668 \text{ L (calc)}$$

$$= 78.0 \text{ L (corr)}$$

12.129 $\quad T = 125°C + 273 = 398 \text{ K} \quad n = \dfrac{3.00 \text{ atm} \times 10.0 \text{ L}}{0.08206 \dfrac{\text{atm L}}{\text{mole K}} \times 398 \text{ K}} = 0.9185581 \text{ (calc)}$

$$= 0.919 \text{ mole } C_2H_2 \text{ (corr)}$$

$$T = 20°C + 273 = 293 \text{ K} \quad n = \frac{0.750 \text{ atm} \times 2.50 \text{ L}}{0.08206 \dfrac{\text{atm L}}{\text{mole K}} \times 293 \text{ K}} = 0.077983 \text{ (calc)}$$

$$= 0.0780 \text{ mole } O_2 \text{ (corr)}$$

$$0.0780 \text{ mole O}_2 \times \frac{4 \text{ moles CO}_2}{5 \text{ moles O}_2} = 0.0624 \text{ mole CO}_2 \text{ (calc and corr)}$$

$$0.919 \text{ mole C}_2\text{H}_2 \times \frac{4 \text{ moles CO}_2}{2 \text{ moles C}_2\text{H}_2} = 1.838 \text{ (calc)} = 1.84 \text{ moles CO}_2 \text{ (corr)}$$

∴ 0.0780 mole O_2 is limiting, and 0.0624 mole CO_2 is produced. $T = 75°C + 273 = 348$ K

$$\text{Volume CO}_2 = \frac{0.0624 \text{ mole} \times 0.08206 \frac{\text{atm L}}{\text{mole K}} \times 348 \text{ K}}{2.00 \text{ atm}} = 0.890974656 \text{ L (calc)} = 0.891 \text{ L (corr)}$$

MIXTURES OF GASES (Sec. 12.15)

12.131 Moles of gas = 3.00 moles N_2 + 3.00 moles O_2 = 6.00 moles gas mixture
$T = 27°C + 273 = 300$ K

$$V = \frac{nRT}{P} = \frac{6.00 \text{ mole} \times 0.08206 \frac{\text{atm L}}{\text{mole K}} \times 300 \text{ K}}{20.00 \text{ atm}} = 7.3854 \text{ (calc)} = 7.39 \text{ L (corr)}$$

12.133 $T = 27°C + 273 = 300$ K

a) $3.00 \text{ g Ne} \times \frac{1 \text{ mole Ne}}{20.18 \text{ g Ne}} = 0.1486620416 \text{ (calc)} = 0.149 \text{ mole Ne (corr)}$

$3.00 \text{ g Ar} \times \frac{1 \text{ mole Ar}}{39.95 \text{ g Ar}} = 0.07509386733 \text{ (calc)} = 0.0751 \text{ mole Ar (corr)}$

Total moles of gas = 0.149 + 0.0751 = 0.2241 (calc) = 0.224 mole (corr)

$$V = \frac{nRT}{P} = \frac{0.224 \text{ mole} \times 0.08206 \frac{\text{atm L}}{\text{mole K}} \times 300 \text{ K}}{1.00 \text{ atm}} = 5.514432 \text{ (calc)} = 5.51 \text{ L (corr)}$$

b) $4.00 \text{ g Ne} \times \frac{1 \text{ mole Ne}}{20.18 \text{ g Ne}} = 0.1982160555 \text{ (calc)} = 0.198 \text{ mole Ne (corr)}$

$2.00 \text{ g Ar} \times \frac{1 \text{ mole Ar}}{39.95 \text{ g Ar}} = 0.05006257822 \text{ (calc)} = 0.0501 \text{ mole Ar (corr)}$

Total moles of gas = 0.198 + 0.0501 = 0.2481 (calc) = 0.248 mole (corr)

$$V = \frac{nRT}{P} = \frac{0.248 \text{ mole} \times 0.08206 \frac{\text{atm L}}{\text{mole K}} \times 300 \text{ K}}{1.00 \text{ atm}} = 6.105264 \text{ (calc)} = 6.11 \text{ L (corr)}$$

c) $5.00 \text{ g Ar} \times \frac{1 \text{ mole Ar}}{39.95 \text{ g Ar}} = 0.1251564456 \text{ (calc)} = 0.125 \text{ mole Ar (corr)}$

Total moles of gas = 3.00 + 0.125 = 3.125 (calc) = 3.12 moles (corr)

$$V = \frac{nRT}{P} = \frac{3.12 \text{ moles} \times 0.08206 \frac{\text{atm L}}{\text{mole K}} \times 300 \text{ K}}{1.00 \text{ atm}} = 76.80816 \text{ (calc)} = 76.8 \text{ L (corr)}$$

d) Total moles of gas $= 4.00 + 4.00 = 8$ (calc) $= 8.00$ moles (corr)

$$V = \frac{nRT}{P} = \frac{8.00 \text{ moles} \times 0.08206 \frac{\text{atm L}}{\text{mole K}} \times 300 \text{ K}}{1.00 \text{ atm}} = 196.944 \text{ (calc)} = 197 \text{ L (corr)}$$

12.135 Determine the pressure (atm) of a 27.0 L gaseous mixture at 20°C if it contains:

a) 3.00 moles each of Ar, Ne, and N_2

$T = 20°C + 273 = 293$ K

Total moles of gas in mixture

$= 3.00$ moles Ar $+ 3.00$ moles Ne $+ 3.00$ moles $N_2 = 9.00$ moles gas

$$P = \frac{nRT}{V} = \frac{9.00 \text{ moles} \times \frac{0.08206 \text{ atm L}}{\text{mole K}} \times 293 \text{ K}}{27.0 \text{ L}} = 8.014526667 \text{ (calc)}$$

$$= 8.01 \text{ atm (corr)}$$

b) 3.00 grams each of Ar, Ne, and N_2

$T = 20°C + 273 = 293$ K

Moles of each gas in the mixture:

$$3.00 \text{ g Ar} \times \frac{1 \text{ mole Ar}}{39.95 \text{ g Ar}} = 0.075093867 \text{ (calc)} = 0.0751 \text{ moles Ar (corr)}$$

$$3.00 \text{ g Ne} \times \frac{1 \text{ mole Ne}}{20.18 \text{ g Ne}} = 0.148662041 \text{ (calc)} = 0.149 \text{ moles Ne (corr)}$$

$$3.00 \text{ g N}_2 \times \frac{1 \text{ mole N}_2}{28.02 \text{ g N}_2} = 0.107066381 \text{ (calc)} = 0.107 \text{ moles N}_2 \text{ (corr)}$$

Total moles of gas in mixture

$= 0.0751$ mole Ar $+ 0.149$ mole Ne $+ 0.107$ mole $N_2 = 0.331$ mole gas

$$P = \frac{nRT}{V} = \frac{0.331 \text{ mole} \times \frac{0.08206 \text{ atm L}}{\text{mole K}} \times 293 \text{ K}}{27.0 \text{ L}} = 0.29475648 \text{ (calc)} = 0.295 \text{ atm (corr)}$$

DALTON'S LAW OF PARTIAL PRESSURES (SEC. 12.16)

12.137 $P_{He} = 9.0$ atm

$P_{Ne} = 14.0$ atm $- 9.0$ atm $= 5$ (corr) $= 5.0$ atm (corr)

$P_{Ar} = 29.0$ atm $- 14.0$ atm $= 15$ (calc) $= 15.0$ atm (corr)

12.139 a) $P_{CO_2} = 842 - 675 = 167$ mm Hg (calc and corr)

b) $P_{N_2} =$ same as before $= 354$ mm Hg (calc and corr)

c) $P_{Ar} =$ same as before $= 235$ mm Hg (calc and corr)

d) $P_{H_2} =$ same as before $= 675 - P_{N_2} - P_{Ar} = 675 - 354 - 235 = 86$ mm Hg (calc and corr)

12.141 Total spheres $= 5$ Ne $+ 3$ Ar $+ 2$ Kr $= 10$ spheres

a) $P_{Ne} = \frac{5 \text{ spheres}}{10 \text{ spheres}} \times 0.060$ atm $= 0.030$ atm (calc and corr)

b) $P_{Ar} = \frac{3 \text{ spheres}}{10 \text{ spheres}} \times 0.060$ atm $= 0.018$ atm (calc and corr)

c) $P_{Kr} = \dfrac{2 \text{ spheres}}{10 \text{ spheres}} \times 0.060 \text{ atm} = 0.012 \text{ atm (calc and corr)}$

12.143 a) moles CO $= 25.0 \text{ g CO} \times \dfrac{1 \text{ mole CO}}{28.01 \text{ g CO}} = 0.8925383792 \text{ (calc)} = 0.893 \text{ mole CO (corr)}$

moles $CO_2 = 25.0 \text{ g } CO_2 \times \dfrac{1 \text{ mole } CO_2}{44.01 \text{ g } CO_2} = 0.5680527153 \text{ (calc)} = 0.568 \text{ mole } CO_2 \text{ (corr)}$

moles $H_2S = 25.0 \text{ g } H_2S \times \dfrac{1 \text{ mole } H_2S}{34.08 \text{ g } H_2S} = 0.733568075 \text{ (calc)} = 0.733 \text{ mole } H_2S \text{ (corr)}$

Total moles $= 0.893 + 0.568 + 0.733 = 2.194$ moles total (calc and corr)

Mole fraction CO $= \dfrac{0.893}{2.194} = 0.4070191431 \text{ (calc)} = 0.407 \text{ mole fraction CO (corr)}$

Mole fraction $CO_2 = \dfrac{0.568}{2.194} = 0.258887876 \text{ (calc)} = 0.259 \text{ mole fraction } CO_2 \text{ (corr)}$

Mole fraction $H_2S = \dfrac{0.733}{2.194} = 0.3340929809 \text{ (calc)} = 0.334 \text{ mole fraction } H_2S \text{ (corr)}$

b) Partial pressure $=$ mole fraction $\times$ total pressure
$P_{CO} = 0.407 \times 1.72 \text{ atm} = 0.70004 \text{ (calc)} = 0.700 \text{ atm (corr)}$
$P_{CO_2} = 0.259 \times 1.72 \text{ atm} = 0.44548 \text{ (calc)} = 0.445 \text{ atm (corr)}$
$P_{H_2S} = 0.334 \times 1.72 \text{ atm} = 0.57448 \text{ (calc)} = 0.574 \text{ atm (corr)}$

12.145 Since mole N_2 + mole $O_2 = 1.00$ mole total, $X_{N_2} + X_{O_2} = 1$

$X_{N_2} =$ mole fraction $N_2 = \dfrac{\text{moles } N_2}{1 \text{ mole total}}$, mole fraction $N_2 =$ moles N_2 present

$X_{O_2} =$ mole fraction $O_2 = \dfrac{\text{moles } O_2}{1 \text{ mole total}}$, mole fraction $O_2 =$ moles O_2 present

	grams N_2	grams O_2	X_{N_2}	X_{O_2}
	14.0 g	16.0 g	0.500	0.500
a)	2.80 g	28.8 g	0.100	0.900
b)	7.00 g	24.0 g	0.250	0.750
c)	2.64 g	29.0 g	0.0942	0.906
d)	21.8 g	7.14 g	0.777	0.223

a) $X_{O_2} =$ moles O_2 present $= 0.900$

$0.900 \text{ mole } O_2 \times \dfrac{32.00 \text{ g } O_2}{1 \text{ mole } O_2} = 28.8 \text{ g } O_2 \text{ (calc and corr)}$

b) $24.0 \text{ g } O_2 \times \dfrac{1 \text{ mole } O_2}{32.00 \text{ g } O_2} = 0.75 \text{ (calc)} = 0.750 \text{ mole } O_2 \text{ (corr)}$

$X_{O_2} =$ moles O_2 present $= 0.750$

$0.250 \text{ mole } N_2 \times \dfrac{28.02 \text{ g } N_2}{1 \text{ mole } N_2} = 7.005 \text{ (calc)} = 7.00 \text{ g } N_2 \text{ (corr)}$

c) $2.64 \text{ g N}_2 \times \dfrac{1 \text{ mole N}_2}{28.02 \text{ g N}_2} = 0.094218415 \text{ (calc)} = 0.0942 \text{ mole N}_2 \text{ (corr)}$

$X_{N_2} = \text{moles N}_2 \text{ present} = 0.0942$

$29.0 \text{ g O}_2 \times \dfrac{1 \text{ mole O}_2}{32.00 \text{ g O}_2} = 0.90625 \text{ (calc)} = 0.906 \text{ mole O}_2 \text{ (corr)}$

$X_{O_2} = \text{moles O}_2 \text{ present} = 0.906 \text{ mole O}_2$

d) $X_{O_2} = \text{moles O}_2 \text{ present} = 0.223 \text{ mole O}_2$

$0.223 \text{ mole O}_2 \times \dfrac{32.00 \text{ g O}_2}{1 \text{ mole O}_2} = 7.136 \text{ (calc)} = 7.14 \text{ g O}_2 \text{ (corr)}$

$X_{N_2} = 1 - 0.223 = 0.777 \text{ (calc and corr)}$

$X_{N_2} = \text{moles N}_2 \text{ present} = 0.777 \text{ mole N}_2$

$0.777 \text{ mole N}_2 \times \dfrac{28.02 \text{ g N}_2}{1 \text{ mole N}_2} = 21.77154 \text{ (calc)} = 21.8 \text{ g N}_2 \text{ (corr)}$

12.147 $P_{He} + P_{Ne} + P_{Ar} = 3.00 \text{ atm}$, partial pressure $=$ mole fraction $\times$ total pressure
Determine the partial pressure of each under each of the following conditions:

a) moles of He $=$ moles Ne $=$ moles Ar; therefore $X_{\text{each gas}} = 1/3$
$P_{\text{each gas}} = 1/3 \times 3.00 \text{ atm} = 1.00 \text{ atm (calc and corr)}$

b) atoms of He $=$ 3 times that of each of the other gases;
number of atoms is proportional to number of moles and mole fraction of each gas:
$3X + X + X = 1.00, 5X = 1, X = 0.2 \text{ (calc)} = 0.200 \text{ (corr)}$
$P_{He} = (3 \times 0.200) \times 3.00 \text{ atm} = 1.80 \text{ atm}$
$P_{Ne} = 0.200 \times 3.00 \text{ atm} = 0.600 \text{ atm}$
$P_{Ar} = 0.200 \times 3.00 \text{ atm} = 0.600 \text{ atm}$

c) partial pressures of Ar, He, and Ne are in a 2:2:3 ratio.
Let $X = P_{\text{each gas}}, 2X + 2X + 3X = P_{\text{total}}, 7X = 3.00 \text{ atm}$,
$X = 0.428571428 \text{ (calc)} = 0.429 \text{ atm (corr)}$
$P_{Ar} = 2 \times 0.429 \text{ atm} = 0.858 \text{ atm}$
$P_{He} = 2 \times 0.429 \text{ atm} = 0.858 \text{ atm}$
$P_{Ne} = 3 \times 0.429 \text{ atm} = 1.287 \text{ (calc)} = 1.29 \text{ atm (corr)}$

d) partial pressure of He $= 1/2$ that of the other two gases. $X = P_{\text{each gas}}$,
$1/2 X + X + X = \text{total pressure } 2.5X = 3.00 \text{ atm}, X = 1.2 \text{ (calc)} = 1.20 \text{ (corr)}$
$P_{He} = 1/2 \times 1.20 \text{ atm} = 0.600 \text{ atm}$
$P_{Ne} = 1.20 \text{ atm}$
$P_{Ar} = 1.20 \text{ atm}$

12.149 $P_A = P_T \cdot \dfrac{n_A}{n_T}$ or $P_T = P_A \cdot \dfrac{n_T}{n_A}$ $P_T = 0.40 \text{ atm} \times \dfrac{4.0 + 2.0 + 0.50 \text{ moles total}}{0.5 \text{ mole Ne}}$

$= 5.2 \text{ atm (calc and corr)}$

12.151 a) $1.00 - 0.150 = 0.850 \text{ mole fraction of O}_2$
$0.850 \times 6.00 \text{ atm} = 5.0 \text{ (calc)} = 5.10 \text{ atm (corr)}$

b) $2.26 \text{ atm} = 0.180 (X \text{ atm})$, thus $X = 12.55555556 \text{ (calc)} = 12.6 \text{ atm total pressure (corr)}$
$1.00 - 0.180 = 0.82 \text{ mole fraction O}_2$
$0.82 (12.6 \text{ atm}) = 10.332 \text{ (calc)} = 10.3 \text{ atm partial pressure of O}_2 \text{ gas (corr)}$

12.153 0.500 atm He + 0.250 atm Ar + 0.350 atm Xe = 1.100 atm total pressure

$$X_{He} = \frac{0.500 \text{ atm}}{1.100 \text{ atm}} = 0.454545 \text{ (calc)} = 0.455 \text{ (corr)}$$

$$X_{Ar} = \frac{0.250 \text{ atm}}{1.100 \text{ atm}} = 0.227272 \text{ (calc)} = 0.227 \text{ (corr)}$$

$$X_{xe} = \frac{0.350 \text{ atm}}{1.100 \text{ atm}} = 0.3181818 \text{ (calc)} = 0.318 \text{ (corr)}$$

12.155 $P_{total} (6.00 \text{ atm}) = P_{He} + P_{CO_2}$ and $P_{He} = X_{He}P_{total}$. $P_{CO_2} = X_{CO_2}P_{total}$
$X_{He} + X_{CO_2} = 1$

	P_{He}	P_{CO_2}	X_{He}	X_{CO_2}
	3.00 atm	3.00 atm	0.500	0.500
a)	4.20 atm	1.80 atm	0.700	0.300
b)	1.80 atm	4.20 atm	0.300	0.700
c)	3.07 atm	2.93 atm	0.511	0.489
d)	2.70 atm	3.30 atm	0.450	0.550

a) $P_{He} = P_{total} - P_{CO_2}$, 6.00 atm $-$ 1.80 atm $= 4.2$ (calc) $= 4.20$ atm (corr)

b) $X_{He} = \dfrac{P_{He}}{P_{total}} = \dfrac{1.80 \text{ atm}}{6.00 \text{ atm}} = 0.3$ (calc) $= 0.300$ (corr)
   $X_{CO_2} = 1 - 0.300 = 0.7$ (calc) $= 0.700$ (corr)

c) $P_{He} = 0.511 \times 6.00 \text{ atm} = 3.066$ (calc) $= 3.07$ atm (corr)
   $P_{CO_2} = 6.00 \text{ atm} - 3.07 \text{ atm} = 2.93$ atm (calc and corr)

d) $P_{CO_2} = 6.00 \text{ atm} - 2.70 \text{ atm} = 3.30$ atm (calc and corr)

$$X_{He} = \frac{P_{He}}{P_{total}} = \frac{2.70 \text{ atm}}{6.00 \text{ atm}} = 0.45 \text{ (calc)} = 0.450 \text{ (corr)}$$

$$X_{CO_2} = 1 - 0.450 = 0.55 \text{ (calc)} = 0.550 \text{ (corr)}$$

12.157 Moles N$_2$ = 20.0 g N$_2$ $\times \dfrac{1 \text{ mole N}_2}{28.02 \text{ g N}_2} = 0.71377588$ (calc) $= 0.714$ mole N$_2$ (corr)

Moles Ar = 8.00 g Ar $\times \dfrac{1 \text{ mole Ar}}{39.95 \text{ g Ar}} = 0.200250$ (calc) $= 0.200$ mole Ar (corr)

Total moles = 0.714 + 0.200 = 0.914 mole

$$\frac{P_1}{n_1} = \frac{P_2}{n_2} \qquad \frac{1.00 \text{ atm}}{0.714 \text{ mole}} = \frac{P_2}{0.914 \text{ mole}} \qquad P_2 = 1.280112 \text{ (calc)} = 1.28 \text{ atm (corr)}$$

12.159 From the coefficients in the balanced equation, the mole fraction of N$_2$ = $\dfrac{1}{1 + 3} = 0.250$ and the

mole fraction of H$_2$ = $\dfrac{3}{1 + 3} = 0.750$.

$P_{N_2} = 0.250 \times 852 \text{ mm Hg} = 213 \text{ mm Hg}$ (calc and corr)
$P_{H_2} = 0.750 \times 852 \text{ mm Hg} = 639 \text{ mm Hg}$ (calc and corr)

12.161 Water vapor pressure values are obtained from Table 12.6 in the text.

a) $743 - 16.5$ mm Hg $= 726.5$ (calc) $= 726$ mm Hg (corr)

b) $645 - 28.3$ mm Hg $= 616.7$ (calc) $= 617$ mm Hg (corr)

c) $762 - 39.9$ mm Hg $= 722.1$ (calc) $= 722$ mm Hg (corr)

d) $0.933 \text{ atm} \times \dfrac{760 \text{ mm Hg}}{1 \text{ atm}} = 709.08$ (calc) $= 709$ mm Hg (corr)

$709 - 18.7$ mm Hg $= 690.3$ (calc) $= 690$ mm Hg (corr)

12.163 a) mole fraction He, $X_{He} = \dfrac{0.100 \text{ mole He}}{0.100 + 0.200 + 0.800 \text{ moles total}}$

$= 0.09090909$ (calc) $= 0.0909$ (corr)

b) mole% Ar $= \dfrac{0.800 \text{ mole Ar}}{1.100 \text{ moles total}} \times 100 = 72.7272727$ (calc) $= 72.7\%$ Ar (corr)

c) $P_{Ne} = X_{Ne} \times P_{total} = \dfrac{0.200 \text{ mole}}{1.100 \text{ moles total}} \times 1.00 \text{ atm} = 0.1818182$ (calc) $= 0.182$ atm (corr)

pressure % Ne $= \dfrac{P_{Ne}}{P_{total}} \times 100 = \dfrac{0.182 \text{ atm}}{1.00 \text{ atm}} \times 100 = 18.2\%$ Ne

d) volume He at STP (alone) $= 0.100 \text{ mole He} \times \dfrac{22.41 \text{ L He}}{1 \text{ mole He}} = 2.241$ (calc) $= 2.24$ L He (corr)

volume % He $= \dfrac{\text{vol He}}{\text{total volume}} \times 100 = \dfrac{2.24 \text{ L He}}{24.64 \text{ L total}} \times 100$

$= 9.0909091$ (calc) $= 9.09\%$ He (corr)

12.165 a) at constant temperature and pressure,

volume % $O_2 = \dfrac{V_{O_2}}{V_{total}} \times 100 = \dfrac{2.0 \text{ L } O_2}{8.0 \text{ L total}} \times 100 = 25\% \ O_2$ (calc and corr)

b) mole Ar $= 3.0 \text{ L Ar} \times \dfrac{1 \text{ mole Ar}}{22.41 \text{ L Ar}} = 0.1338688086$ (calc) $= 0.13$ mole Ar (corr)

mole $O_2 = 2.0 \text{ L } O_2 \times \dfrac{1 \text{ mole } O_2}{22.41 \text{ L } O_2} = 0.08924587238$ (calc) $= 0.089$ mole O_2 (corr)

mole Ne $= 3.0 \text{ L Ne} \times \dfrac{1 \text{ mole Ne}}{22.41 \text{ L Ne}} = 0.1338688086$ (calc) $= 0.13$ mole Ne (corr)

total moles $= 0.13 + 0.089 + 0.13 = 0.349$ (calc) $= 0.35$ mole (corr)

mole % Ar $= \dfrac{\text{moles Ar}}{\text{total moles}} \times 100 = \dfrac{0.13}{0.35} \times 100 = 37.142857$ (calc) $= 37\%$ Ar (corr)

c) In the final container, $P_{Ne} = X_{Ne} \times P_{total} = \dfrac{0.13 \text{ mole Ne}}{0.35 \text{ mole total}} \times 1.00 \text{ atm}$

$= 0.37142857$ (calc) $= 0.37$ atm (corr)

pressure % Ne $= \dfrac{0.37 \text{ atm Ne}}{1.0 \text{ atm total}} \times 100 = 37\%$ Ne (calc and corr)

d) $P_{O_2} = X_{O_2} \times P_{total}$ in final container $= \dfrac{0.089 \text{ mole } O_2}{0.35 \text{ mole total}} \times 1.0 \text{ atm}$

$= 0.25428571 \text{ (calc)} = 0.25 \text{ atm (corr)}$

Multi-Concept Problems

12.167 Increasing density of gases at STP: $F_2 < CO_2 < NO_2 < SO_2 < Cl_2$

12.169 All gases have the same molar volume at STP. $O_2 = NO = CO = N_2 = NH_3$

12.171 $1.00 \text{ L } CO_2 \times \dfrac{1 \text{ mole } CO_2}{22.41 \text{ L } CO_2} \times \dfrac{6.022 \times 10^{23} \text{ molecules } CO_2}{1 \text{ mole } CO_2}$

$= 2.6871932 \times 10^{22} \text{ (calc)} = 2.69 \times 10^{22} \text{ molecules } CO_2 \text{ (corr)}$

12.173 $8.00 \text{ g } N_2 \times \dfrac{1 \text{ mole } N_2}{28.02 \text{ g } N_2} = 0.2855103 \text{ (calc)} = 0.286 \text{ mole } N_2 \text{ (corr)}$

$1.00 \times 10^{23} \text{ molecules } NH_3 \times \dfrac{1 \text{ mole } NH_3}{6.022 \times 10^{23} \text{ molecules } NH_3}$

$= 0.1660577 \text{ (calc)} = 0.166 \text{ mole } NH_3 \text{ (corr)}$

$V_{NH_3} = V_{N_2} \times \dfrac{n_{NH_3}}{n_{H_2}} = 6.00 \text{ L} \times \dfrac{0.166 \text{ mole}}{0.286 \text{ mole}} = 3.4825174 \text{ (calc)} = 3.48 \text{ L } NH_3 \text{ (corr)}$

12.175 $n_1 = \dfrac{1.00 \text{ atm} \times 1.00 \text{ L}}{0.08206 \dfrac{\text{atm L}}{\text{mole K}} \times 298 \text{ K}} = 0.04089330609 \text{ (calc)} = 0.0409 \text{ mole } O_2 \text{ (corr)}$

$n_2 = \dfrac{0.880 \text{ atm} \times 1.00 \text{ L}}{0.08206 \dfrac{\text{atm L}}{\text{mole K}} \times 295 \text{ K}} = 0.0363520698 \text{ (calc)} = 0.0364 \text{ mole } O_2 \text{ (corr)}$

$\Delta n = (0.0409 - 0.0364) \text{ mole } O_2 = 0.0045 \text{ mole } O_2 \text{ (calc and corr)}$

$0.0045 \text{ mole } O_2 \times \dfrac{32.00 \text{ g } O_2}{1 \text{ mole } O_2} = 0.144 \text{ (calc)} = 0.14 \text{ g } O_2 \text{ (corr)}$

12.177 Let V_1 = original volume in mL, $V_2 = V_1 - 25.0 \text{ mL}$ Applying Boyle's law: $V_2 = V_1 \times \dfrac{P_1}{P_2}$

$V_1 - 25.0 = V_1 \times \dfrac{1.50 \text{ atm}}{3.50 \text{ atm}}$ Rearranging: $V_1 - V_1 \times \dfrac{1.50}{3.50} = 25.0$ or $V_1\left(1 - \dfrac{1.50}{3.50}\right) = 25.0$

or $V_1 \dfrac{3.50 - 1.50}{3.50} = 25.0$ or $V_1 \times \dfrac{2.00}{3.50} = 25.0$

$V_1 = 25.0 \times \dfrac{3.50}{2.00} = 43.75 \text{ (calc)} = 43.8 \text{ mL (corr)}$

12.179 $P_1 V_1 = P_2 V_2 \qquad V_2 = 0.800\, V_1$

$$P_2 = P_1 \times \frac{V_1}{V_2} = P_1 \frac{V_1}{0.800\, V_1} = 1.25\, P_1 \ (\text{calc and corr})$$

$$\Delta P = (1.25\, P_1 - P_1) = 0.25\, P_1$$

$$\% \text{ change} = \frac{\Delta P}{P_1} \times 100 = \frac{0.25\, P_1}{P_1} \times 100 = 25\% \ (\text{calc and corr})$$

12.181 a) $\dfrac{d_{O_2}}{d_{N_2}} = \dfrac{\dfrac{\text{molar mass}_{O_2}}{\text{molar volume (STP)}}}{\dfrac{\text{molar mass}_{N_2}}{\text{molar volume (STP)}}} = \dfrac{\text{molar mass}_{O_2}}{\text{molar mass}_{N_2}} = \dfrac{32.00 \text{ g/mole}}{28.02 \text{ g/mole}}$

$$= 1.1420414 \ (\text{calc}) = 1.14{:}1 \ (\text{corr})$$

b) $\dfrac{d_{O_2}}{d_{N_2}} = \dfrac{\dfrac{(M)_{O_2} P_{O_2}}{RT}}{\dfrac{(M)_{N_2} P_{N_2}}{RT}} = \dfrac{(M)_{O_2} \times 1.25 \text{ atm}}{(M)_{N_2} \times 1.25 \text{ atm}} = \dfrac{32.00 \text{ g/mole}}{28.02 \text{ g/mole}} = 1.1420414 \ (\text{calc}) = 1.14{:}1 \ (\text{corr})$

12.183 Find the pressure of SF_6 in the new container. This is its partial pressure.

$$P_{SF_6} = 0.97 \text{ atm} \times \frac{376 \text{ mL}}{275 \text{ mL}} = 1.3262545 \ (\text{calc}) = 1.3 \text{ atm} \ (\text{corr})$$

12.185 $P_{Ar} = \dfrac{g\,RT}{V(M)} = \dfrac{15.0 \text{ g} \times 0.08206 \dfrac{\text{atm L}}{\text{mole K}} \times 327 \text{ K}}{4.0 \text{ L} \times 39.95 \text{ g/mole}} = 2.5188004 \ (\text{calc}) = 2.5 \text{ atm} \ (\text{corr})$

12.187 N_2 pressure in He container at 37°C (310 K):

$$P_2 = 645 \text{ mm Hg} \times \frac{30.0 \text{ mL}}{40.0 \text{ mL}} \times \frac{310 \text{ K}}{300 \text{ K}} = 499.875 \ (\text{calc}) = 5\overline{00} \text{ mm Hg} \ (\text{corr})$$

Total pressure in 40.0 mL container at 37°C:

$$P_{total} = 765 + 5\overline{00} \text{ mm Hg} = 1265 \text{ mm Hg} \ (\text{calc and corr})$$

Total pressure at 32°C:

$$P_2 = 1265 \text{ mm Hg} \times \frac{305 \text{ K}}{310 \text{ K}} = 1244.5968 \ (\text{calc}) = 1240 \text{ mm Hg} \ (\text{corr})$$

Answers to Multiple-Choice Practice Test

MC 12.1	e	MC 12.2	c	MC 12.3	d	MC 12.4	e	MC 12.5	b
MC 12.6	b	MC 12.7	e	MC 12.8	a	MC 12.9	e	MC 12.10	b
MC 12.11	d	MC 12.12	c	MC 12.13	d	MC 12.14	d	MC 12.15	c
MC 12.16	a	MC 12.17	b	MC 12.18	a	MC 12.19	d	MC 12.20	c

CHAPTER THIRTEEN
Solutions

Practice Problems by Topic

CHARACTERISTICS OF SOLUTIONS (SEC. 13.1)

13.1   a) True, solutions may have multiple solutes.
        b) True, if it is a solution, it is homogeneous by definition.
        c) True, all parts of a homogeneous solution have the same properties.
        d) False, solutes will not settle out of solution.

13.3   a) solute = sodium chloride, solvent = water
        b) solute = sucrose, solvent = water
        c) solute = water, solvent = ethyl alcohol
        d) solute = ethyl alcohol, solvent = methyl alcohol

SOLUBILITY (SEC. 13.2)

13.5   a) slightly soluble   b) soluble   c) very soluble   d) very soluble

13.7   a) supersaturated   b) saturated   c) unsaturated   d) saturated

13.9   a) $\dfrac{161.4 \text{ g CsCl}}{100 \text{ g water}}$ at 0°C is saturated

        b) $\dfrac{161.4 \text{ g CsCl}}{100 \text{ g water}}$ at 50°C is unsaturated

        c) $\dfrac{161.4 \text{ g CsCl}}{200 \text{ g water}} = \dfrac{80.70 \text{ g CsCl}}{100 \text{ g water}}$ at 50°C is unsaturated

        d) $\dfrac{161.4 \text{ g CsCl}}{50 \text{ g water}} = \dfrac{322.8 \text{ g CsCl}}{100 \text{ g water}}$ at 100°C is supersaturated

13.11   a) $\dfrac{0.50 \text{ g Ag}_2\text{SO}_4}{100 \text{ g water}}$ at 100°C is dilute

        b) $\dfrac{0.50 \text{ g Ag}_2\text{SO}_4}{100 \text{ g water}}$ at 0°C is concentrated

        c) $\dfrac{0.50 \text{ g Ag}_2\text{SO}_4}{50 \text{ g water}} = \dfrac{1.00 \text{ g Ag}_2\text{SO}_4}{100 \text{ g water}}$ at 50°C is concentrated

        d) $\dfrac{0.050 \text{ g Ag}_2\text{SO}_4}{10 \text{ g water}} = \dfrac{0.50 \text{ g Ag}_2\text{SO}_4}{100 \text{ g water}}$ at 0°C is concentrated

13.13 a) Second, ammonia gas in water with $P = 1$ atm and $T = 50°C$ is more soluble.
       b) First, carbon dioxide gas in water with $P = 2$ atm and $T = 75°C$ is more soluble.
       c) Second, table salt in water with $P = 1$ atm and $T = 60°C$ is more soluble.
       d) Second, table sugar in water with $P = 1$ atm and $T = 70°C$ is more soluble.

13.15 $200 \text{ mL H}_2\text{O} \times \dfrac{1 \text{ g H}_2\text{O}}{1 \text{ mL H}_2\text{O}} = 200 \text{ g H}_2\text{O}$ (calc and corr)

      a) At 70°C, solubility $\dfrac{110 \text{ g Pb(NO}_3)_2}{100 \text{ g water}}$ if $\dfrac{200 \text{ g Pb(NO}_3)_2}{200 \text{ g water}}$ present, none will settle out of water.

      b) At 40°C, solubility $\dfrac{78 \text{ g Pb(NO}_3)_2}{100 \text{ g water}}$ if $\dfrac{200 \text{ g Pb(NO}_3)_2}{200 \text{ g water}}$ present,

               then $200 \text{ g} - 2(78 \text{ g}) = 44 \text{ g Pb(NO}_3)_2$ will settle out of water.

SOLUTION FORMATION (SEC. 13.3)

13.17 a) hydrated ion   b) hydrated ion   c) oxygen atom   d) hydrogen atom

13.19 a) decrease   b) increase   c) increase   d) increase

SOLUBILITY RULES (SEC. 13.4)

13.21 a) ethanol   b) carbon tetrachloride   c) ethanol   d) ethanol

13.23 a) Acetate and nitrate ions are soluble.
       b) Sulfates and chlorides are soluble with exceptions.
       c) Sodium and ammonium ions are soluble.
       d) Bromides and sulfates are soluble with exceptions.

13.25 a) Magnesium nitrate, $Mg(NO_3)_2$, all nitrates are soluble in water (no exceptions).
       b) Calcium acetate, $Ca(C_2H_3O_2)_2$, all acetates are soluble in water (no exceptions).
       c) $NiCl_2$, nickel(II) chloride, chlorides, bromides, and iodides are soluble in water (unless combined
          with Ag^+, Pb^{2+}, or Hg_2^{2+} ions).
       d) $Al_2(SO_4)_3$, aluminum sulfate, sulfates are soluble in water (unless combined with Ca^{2+}, Sr^{2+}, Ba^{2+},
          Pb^{2+}, or Ag^+ ions).

13.27 a) Calcium carbonate, $CaCO_3$, insoluble in water; carbonates are insoluble (unless combined with
          group IA or NH_4^+ ions).
       b) Aluminum sulfide, Al_2S_3, insoluble in water; sulfides are insoluble (unless combined with NH_4^+,
          group IA, Ca^{2+}, Sr^{2+}, or Ba^{2+} ions).
       c) $Cu(OH)_2$, copper(II) hydroxide, insoluble in water; hydroxides are insoluble (unless combined with
          group IA, Ca^{2+}, Sr^{2+}, Ba^{2+} ions).
       d) $PbSO_4$, lead(II) sulfate, insoluble in water; sulfates are soluble (unless combined with Ca^{2+}, Sr^{2+},
          Ba^{2+}, Pb^{2+}, or Ag^+ ions).

13.29 a) both soluble   b) both soluble   c) insoluble and soluble   d) both insoluble

13.31 a) Magnesium sulfate, $MgSO_4$, is soluble in water.
         Calcium sulfate, $CaSO_4$, is insoluble in water.
       b) Magnesium carbonate, $MgCO_3$, is insoluble in water.
         Calcium carbonate, $CaCO_3$, is insoluble in water.

    c) Magnesium nitrate, $Mg(NO_3)_2$, is soluble in water.
       Calcium nitrate, $Ca(NO_3)_2$, is soluble in water.
    d) Magnesium hydroxide, $Mg(OH)_2$, is insoluble in water.
       Calcium hydroxide, $Ca(OH)_2$, is soluble in water.

MASS PERCENT (SEC. 13.4)

13.33   Mass percent of NaCl in each solution:

$$\text{mass percent (NaCl)} = \frac{\text{mass NaCl}}{\text{mass NaCl} + \text{mass } H_2O} \times 100$$

  a) 10.0 g NaCl, 40.0 g H_2O

$$\text{mass percent (NaCl)} = \frac{10.0 \text{ g NaCl}}{10.0 \text{ g NaCl} + 40.0 \text{ g } H_2O} \times 100 = 20 \text{ (calc)} = 20.0\% \text{ (corr)}$$

  b) 5.00 g NaCl, 355 g H_2O

$$\text{mass percent (NaCl)} = \frac{5.00 \text{ g NaCl}}{5.00 \text{ g NaCl} + 355 \text{ g } H_2O} \times 100$$

$$= 1.3888888889 \text{ (calc)} = 1.39\% \text{ (corr)}$$

  c) 20.0 g NaCl, 565.0 g H_2O

$$\text{mass percent (NaCl)} = \frac{20.0 \text{ g NaCl}}{20.0 \text{ g NaCl} + 565.0 \text{ g } H_2O} \times 100$$

$$= 3.418803419 \text{ (calc)} = 3.42\% \text{ (corr)}$$

  d) 529 mg NaCl, 2.00 g H_2O $529 \text{ mg} \times \dfrac{10^{-3} \text{ g}}{1 \text{ mg}} = 0.529 \text{ g (calc and corr)}$

$$\text{mass percent (NaCl)} = \frac{0.529 \text{ g NaCl}}{0.529 \text{ g NaCl} + 2.00 \text{ g } H_2O} \times 100$$

$$= 20.91735864 \text{ (calc)} = 20.9\% \text{ (corr)}$$

13.35   Mass percent of $NaNO_3$ in each solution:

mass solution − mass solvent = mass solute

$$\text{mass percent (NaNO}_3) = \frac{\text{mass NaNO}_3}{\text{mass solution}} \times 100$$

  a) 35.0 g H_2O, 38.0 g solution

    mass solute = 38.0 g solution − 35.0 g H_2O = 3.0 g solute

$$\text{mass percent (NaNO}_3) = \frac{3.0 \text{ g NaNO}_3}{38.0 \text{ g solution}} \times 100 = 7.894736842 \text{ (calc)} = 7.9\% \text{ (corr)}$$

  b) 36.0 g H_2O, 52.0 g solution

    mass solute = 52.0 g solution − 36.0 g H_2O = 16.0 g solute

$$\text{mass percent (NaNO}_3) = \frac{16.0 \text{ g NaNO}_3}{52.0 \text{ g solution}} \times 100 = 30.76923077 \text{ (calc)} = 30.8\% \text{ (corr)}$$

c) 45.0 g H_2O, 45.9 g solution

   mass solute = 45.9 g solution − 45.0 g H_2O = 0.9 g solute

   mass percent ($NaNO_3$) = $\dfrac{0.9 \text{ g } NaNO_3}{45.9 \text{ g solution}}$ × 100 = 1.960784314 (calc) = 2 % (corr)

d) 623 g H_2O, 677 g solution

   mass solute = 677 g solution − 623 g H_2O = 54 g solute

   mass percent ($NaNO_3$) = $\dfrac{54 \text{ g } NaNO_3}{677 \text{ g solution}}$ × 100 = 7.976366322 (calc) = 8.0 % (corr)

13.37   Mass percent of Na_2SO_4 can be defined as: $\dfrac{\text{mass } Na_2SO_4}{100 \text{ g solution}}$

   mass solute = mass solution × $\dfrac{\text{mass } Na_2SO_4}{100 \text{ g solution}}$

a) 45.0 g solution, 3.00 % (m/m) Na_2SO_4

   45.0 g solution × $\dfrac{3.00 \text{ g } Na_2SO_4}{100 \text{ g solution}}$ = 1.35 g Na_2SO_4 (calc and corr)

b) 575 g solution, 15.0 % (m/m) Na_2SO_4

   575 g solution × $\dfrac{15.0 \text{ g } Na_2SO_4}{100 \text{ g solution}}$ = 86.25 (calc) = 86.2 g Na_2SO_4 (corr)

c) 2.00 g solution, 0.100 % (m/m) Na_2SO_4

   2.00 g solution × $\dfrac{0.100 \text{ g } Na_2SO_4}{100 \text{ g solution}}$ = 2×10^{-3} (calc) = 2.00×10^{-3} g Na_2SO_4 (corr)

d) 66.6 g solution, 30.0 % (m/m) Na_2SO_4

   66.6 g solution × $\dfrac{30.0 \text{ g } Na_2SO_4}{100 \text{ g solution}}$ = 19.98 (calc) = 20.0 g Na_2SO_4 (corr)

13.39

	Solute mass	Solvent mass	Solution mass	Mass % solute
	10.0 g	20.0 g	30.0 g	33.3 %
a)	12.0 g	22.0 g	34.0 g	35.3 %
b)	24.0 g	50.0 g	74.0 g	32.4 %
c)	36.0 g	40.0 g	76.0 g	47.4 %
d)	40.0 g	52.0 g	92.0 g	43.5 %

a) mass solvent = 34.0 g solution − 12.0 g solute = 22.0 g solvent (calc and corr)

   % (m/m) solute = $\dfrac{12.0 \text{ g solute}}{34.0 \text{ g solution}}$ × 100 = 35.29411765 (calc) = 35.3 % (corr)

b) mass solution = 24.0 g solute + 50.0 g solvent = 74.0 g solution

$$\%(m/m) \text{ solute} = \frac{24.0 \text{ g solute}}{74.0 \text{ g solution}} \times 100 = 32.43243243 \text{ (calc)} = 32.4 \% \text{ (corr)}$$

c) mass percent of solute = 47.4 %,
100 % − 47.4 % = 52.6 % mass percent of solvent
100 g solution contains 47.4 g solute and 52.6 g solvent

$$\text{solute mass} = 40.0 \text{ g solvent} \times \frac{47.4 \text{ g solute}}{52.6 \text{ g solvent}} = 36.04562738 \text{ (calc)} = 36.0 \text{ g solute (corr)}$$

mass solution = 36.0 g solute + 40.0 solvent = 76.0 g solution

d) $$\text{mass solute} = 92.0 \text{ g solution} \times \frac{43.5 \text{ g solute}}{100 \text{ g solution}} = 40.02 \text{ (calc)} = 40.0 \text{ g solute (corr)}$$

mass solvent = 92.0 g solution − 40.0 g solute = 52.0 g solvent

13.41 a) $$5.75 \text{ g solution} \times \frac{10.00 \text{ g CaCl}_2}{100.0 \text{ g solution}} = 0.575 \text{ g CaCl}_2 \text{ (calc and corr)}$$

5.75 g solution − 0.575 g CaCl$_2$ = 5.175 (calc) = 5.18 g H$_2$O (corr)

Alternate method: If the solution is 10.00 % CaCl$_2$, it is 90.00 % H$_2$O.

$$5.75 \text{ g solution} \times \frac{90.00 \text{ g H}_2\text{O}}{100.0 \text{ g solution}} = 5.175 \text{ (calc)} = 5.18 \text{ g H}_2\text{O (corr)}$$

b) $$57.5 \text{ g solution} \times \frac{10.00 \text{ g CaCl}_2}{100.0 \text{ g solution}} = 5.75 \text{ g CaCl}_2 \text{ (calc and corr)}$$

57.5 g solution − 5.75 g CaCl$_2$ = 51.75 g (calc) = 51.8 g H$_2$O (corr)

c) $$57.5 \text{ g solution} \times \frac{1.00 \text{ g CaCl}_2}{100 \text{ g solution}} = 0.575 \text{ g CaCl}_2 \text{ (calc and corr)}$$

57.5 g solution − 0.575 g CaCl$_2$ = 56.925 (calc) = 56.9 g H$_2$O (corr)

d) $$2.3 \text{ g solution} \times \frac{0.80 \text{ g CuCl}_2}{100 \text{ g solution}} = 0.0184 \text{ (calc)} = 0.018 \text{ g CaCl}_2 \text{ (corr)}$$

2.3 g solution − 0.018 g CaCl$_2$ = 2.282 (calc) = 2.3 g H$_2$O (corr)

13.43 a) $$50.0 \text{ g NaCl} \times \frac{95.00 \text{ g H}_2\text{O}}{5.00 \text{ g NaCl}} = 950 \text{ (calc)} = 9.50 \times 10^2 \text{ g H}_2\text{O (corr)}$$

b) $$50.0 \text{ g KCl} \times \frac{95.00 \text{ g H}_2\text{O}}{5.00 \text{ g KCl}} = 950 \text{ (calc)} = 9.50 \times 10^2 \text{ g H}_2\text{O (corr)}$$

c) $$50.0 \text{ g Na}_2\text{SO}_4 \times \frac{95.00 \text{ g H}_2\text{O}}{5.00 \text{ g Na}_2\text{SO}_4} = 950 \text{ (calc)} = 9.50 \times 10^2 \text{ g H}_2\text{O (corr)}$$

d) $$50.0 \text{ g LiNO}_3 \times \frac{95.00 \text{ g H}_2\text{O}}{5.00 \text{ g LiNO}_3} = 950 \text{ (calc)} = 9.50 \times 10^2 \text{ g H}_2\text{O (corr)}$$

VOLUME PERCENT (SEC. 13.6)

13.45   volume percent (ethanol) $= \dfrac{\text{volume ethanol}}{\text{volume solution}} \times 100$

    a) volume percent ethanol $= \dfrac{257 \text{ mL ethanol}}{325 \text{ mL solution}} \times 100 = 79.07692308 \text{ (calc)} = 79.1 \text{ \% (corr)}$

    b) $3.25 \text{ L} \times \dfrac{1 \text{ mL}}{10^{-3} \text{ L}} = 3250 \text{ mL}$

        volume percent ethanol $= \dfrac{257 \text{ mL ethanol}}{3250 \text{ mL solution}} \times 100 = 7.907692308 \text{ (calc)} = 7.91 \text{ \% (corr)}$

    c) volume percent ethanol $= \dfrac{20.0 \text{ mL ethanol}}{35.0 \text{ mL solution}} \times 100 = 57.14285714 \text{ (calc)} = 57.1 \text{ \% (corr)}$

    d) volume percent ethanol $= \dfrac{75.0 \text{ L ethanol}}{625 \text{ L solution}} \times 100 = 12 \text{ (calc)} = 12.0 \text{ \% (corr)}$

13.47 a) $\dfrac{360.6 \text{ mL methyl alcohol}}{1000.0 \text{ mL solution}} \times 100 = 36.06 \text{ \% (v/v) methyl alcohol (calc and corr)}$

     b) $\dfrac{667.2 \text{ mL H}_2\text{O}}{1000.0 \text{ mL solution}} \times 100 = 66.72 \text{ \% (v/v) H}_2\text{O (calc and corr)}$

13.49 $225 \text{ mL solution} \times \dfrac{1.25 \text{ mL C}_3\text{H}_8\text{O}}{100 \text{ mL solution}} = 2.8125 \text{ (calc)} = 2.81 \text{ mL C}_3\text{H}_8\text{O (corr)}$

13.51 $4.00 \text{ gal solution} \times \dfrac{35.0 \text{ gal H}_2\text{O}}{100.0 \text{ gal solution}} = 1.4 \text{ (calc)} = 1.40 \text{ gal H}_2\text{O (corr)}$

MASS–VOLUME PERCENT (SEC. 13.6)

13.53   mass-volume percent $= \dfrac{\text{mass solute (g)}}{\text{volume solution (mL)}} \times 100$

    a) mass-volume percent (NaNO$_3$) $= \dfrac{0.325 \text{ g NaNO}_3}{375 \text{ mL solution}} \times 100 = 0.086666667 \text{ (calc)} = 0.0867 \text{ \% (corr)}$

    b) mass-volume percent (NaNO$_3$) $= \dfrac{1.75 \text{ g NaNO}_3}{20.0 \text{ mL solution}} \times 100 = 8.75 \text{ \% (calc and corr)}$

    c) mass-volume percent (NaNO$_3$) $= \dfrac{83.2 \text{ g NaNO}_3}{975 \text{ mL solution}} \times 100 = 8.533333333 \text{ (calc)} = 8.53 \text{ \% (corr)}$

    d) mass-volume percent (NaNO$_3$) $= \dfrac{0.033 \text{ g NaNO}_3}{2.00 \text{ mL solution}} \times 100 = 1.65 \text{ (calc)} = 1.6 \text{ \% (corr)}$

13.55   mass-volume percent $= \dfrac{\text{mass solute (g)}}{\text{volume solution (mL)}} \times 100$

a) mass-volume percent $(NaNO_3)$ $\dfrac{0.445 \text{ g } NaNO_3}{575 \text{ mL solution}} \times 100 = 0.077391304 \text{ (calc)} = 0.0774 \text{ % (corr)}$

b) $20.0 \text{ L} \times \dfrac{1 \text{ mL}}{10^{-3} \text{ L}} = 20.0 \times 10^3 \text{ mL}$

mass-volume percent $(NaNO_3)$ $\dfrac{11.75 \text{ g } NaNO_3}{20.0 \times 10^3 \text{ mL solution}} \times 100$
$= 0.05875 \text{ (calc)} = 0.0588 \text{ % (corr)}$

c) $83.2 \text{ mg} \times \dfrac{10^{-3} \text{ g}}{1 \text{ mg}} = 8.32 \times 10^{-2} \text{ g}$

mass-volume percent $(NaNO_3)$ $\dfrac{8.32 \times 10^{-2} \text{ g } NaNO_3}{9.75 \text{ mL solution}} \times 100$
$= 0.853333333 \times 10^{-3} \text{ (calc)} = 0.853 \times 10^{-3} \text{ % (corr)}$

d) $0.033 \text{ mole } NaNO_3 \times \dfrac{85.00 \text{ g } NaNO_3}{1 \text{ mole } NaNO_3} = 2.805 \text{ (calc)} = 2.8 \text{ g } NaNO_3 \text{ (corr)}$

mass-volume percent $(NaNO_3)$ $\dfrac{2.8 \text{ g } NaNO_3}{20.00 \text{ mL solution}} \times 100 = 14 \text{ % (calc and corr)}$

13.57   a) $5.00 \text{ g NaCl} \times \dfrac{100 \text{ mL solution}}{6.00 \text{ g NaCl}} = 83.33333 \text{ (calc)} = 83.3 \text{ mL solution (corr)}$

b) $7.00 \text{ g NaCl} \times \dfrac{100 \text{ mL solution}}{6.00 \text{ g NaCl}} = 116.666667 \text{ (calc)} = 117 \text{ mL solution (corr)}$

c) $225 \text{ g NaCl} \times \dfrac{100 \text{ mL solution}}{6.00 \text{ g NaCl}} = 3750 \text{ mL solution (calc and corr)}$

d) $225 \text{ mg NaCl} \times \dfrac{10^{-3} \text{ g}}{1 \text{ mg}} \times \dfrac{100 \text{ mL solution}}{6.00 \text{ g NaCl}} = 3.75 \text{ mL solution (calc and corr)}$

13.59   a) $455 \text{ mL solution} \times \dfrac{2.50 \text{ g } Na_3PO_4}{100 \text{ mL solution}} = 11.375 \text{ (calc)} = 11.4 \text{ g } Na_3PO_4 \text{ (corr)}$

b) $50.0 \text{ L solution} \times \dfrac{1 \text{ mL}}{10^{-3} \text{ L}} \times \dfrac{7.50 \text{ g } Na_3PO_4}{100 \text{ mL solution}} = 3750 \text{ (calc)} = 3.75 \times 10^3 \text{ g } Na_3PO_4 \text{ (corr)}$

13.61   Volume of solution $= 5.0 \text{ g CsCl} + 20.0 \text{ g } H_2O = 25.0 \text{ g solution} \times \dfrac{1 \text{ mL solution}}{1.18 \text{ g solution}}$
$= 21.1864407 \text{ (calc)} = 21.2 \text{ mL solution (corr)}$

$\% \left(\dfrac{m}{v}\right) = \dfrac{5.0 \text{ g CsCl}}{21.2 \text{ mL solution}} \times 100 = 23.5849057 \text{ (calc)} = 24\% \text{ (m/v) CsCl (corr)}$

PARTS PER MILLION AND PARTS PER BILLION (SEC. 13.7)

13.63   a) $37.5 \, \mu g \, NaCl \times \dfrac{10^{-3} \, g}{1 \, mg} = 0.0375 \, g \, NaCl$

$21.0 \, kg \, H_2O \times \dfrac{10^3 \, g}{1 \, kg} = 21,\overline{0}00 \, g \, H_2O$

solution mass $= 0.0375 \, g + 21,\overline{0}00 \, g = 21,\overline{0}00 \, g \, solution$

$\dfrac{0.0375 \, g \, NaCl}{21,\overline{0}00 \, g \, solution} \times 10^6 = 1.7857143 \, (calc) = 1.79 \, ppm \, NaCl \, (corr)$

b) $2.12 \, cg \, NaCl \times \dfrac{10^{-2} \, g}{1 \, cg} = 0.0212 \, g \, NaCl$

solution mass $= 0.0212 \, g \, NaCl + 125 \, g \, H_2O = 125.0212 \, (calc) = 125 \, g \, solution \, (corr)$

$\dfrac{0.0212 \, g \, NaCl}{125 \, g \, solution} \times 10^6 = 169.6 \, (calc) = 1\overline{7}0 \, ppm \, NaCl$

c) $1.00 \, \mu g \, NaCl = 1.00 \times 10^{-6} \, g \, NaCl; \quad 32.0 \, dg \, H_2O \times \dfrac{10^{-1} \, g}{1 \, dg} = 3.20 \, g \, H_2O$

solution mass $= 3.20 \, g \, H_2O + 1 \times 10^{-6} \, g \, NaCl = 3.200001 \, (calc) = 3.20 \, g \, solution \, (corr)$

$\dfrac{1.00 \times 10^{-6} \, g \, NaCl}{3.20 \, g \, solution} \times 10^6 = 0.3125 \, (calc) = 0.312 \, ppm \, NaCl \, (corr)$

d) $35.7 \, mg \, NaCl \times \dfrac{10^{-3} \, g}{1 \, mg} = 0.0357 \, g \, NaCl$

solution mass $= 15.7 \, g \, H_2O + 0.0325 \, g \, NaCl = 15.7325 \, (calc) = 15.7 \, g \, solution \, (corr)$

$\dfrac{0.0357 \, g \, NaCl}{15.7 \, g \, solution} \times 10^6 = 2273.88535 \, (calc) = 2,270 \, ppm \, NaCl \, (corr)$

13.65   The only difference in set-up for ppb compared with ppm is changing the factor 10^6 to 10^9.
The ppb value is 1000 times the ppm value.
a) 1790 ppb   b) $17\overline{0},000$ ppb   c) 312 ppb   d) 2,270,000 ppb

13.67 $1.00 \, mL \, SO_2 \, air \times \dfrac{10^6 \, mL \, air}{0.087 \, mL \, SO_2} \times \dfrac{10^{-3} \, L}{1 \, mL} = 11494.25287 \, (calc) = 11,000 \, L \, air \, (corr)$

13.69 $ppm = \dfrac{mass \, solute}{mass \, solution} \times 10^6 \qquad\qquad ppb = \dfrac{mass \, solute}{mass \, solution} \times 10^9$

$pph = \dfrac{mass \, solute}{mass \, solution} \times 10^2 \quad = \quad \% \, m/m = \dfrac{mass \, solute}{mass \, solution} \times 10^2$

	ppm	ppb	pph	% (m/m)
	3.7	3700	0.00037	0.00037
a)	13	13,000	0.0013	0.0013
b)	231	231,000	0.0231	0.0231
c)	0.5	500	0.00005	0.00005
d)	22,000	22,000,000	2.2	2.2

a) ppm → pph $\qquad 13\ \text{ppm} \times \dfrac{10^2\ \text{pph}}{10^6\ \text{ppm}} = 0.0013\ \text{pph (calc and corr)}$

ppm → ppb $\qquad 13\ \text{ppm} \times \dfrac{10^9\ \text{ppb}}{10^6\ \text{ppm}} = 13{,}000\ \text{ppb (calc and corr)}$

b) ppb → ppm $\qquad 231{,}000\ \text{ppb} \times \dfrac{10^6\ \text{ppm}}{10^9\ \text{ppb}} = 231\ \text{ppm (calc and corr)}$

pph = % (m/m) $0.0231\ \text{pph} = 0.0231\ \%\ \text{(m/m) (corr)}$

c) ppm → ppb $\qquad 0.5\ \text{ppm} \times \dfrac{10^9\ \text{ppb}}{10^6\ \text{ppm}} = 500\ \text{ppb (calc and corr)}$

% (m/m) = $\qquad$ pph $0.00005\ \%\ \text{(m/m)} = 0.00005\ \text{pph}$

d) pph = % (m/m) $\qquad 2.2\ \text{pph} = 2.2\ \%\ \text{(m/m)}$

pph → ppm $\quad 2.2\ \text{pph} \times \dfrac{10^6\ \text{ppm}}{10^2\ \text{pph}} = 22{,}000\ \text{ppm (calc and corr)}$

pph → ppb $\quad 2.2\ \text{pph} \times \dfrac{10^9\ \text{ppb}}{10^2\ \text{pph}} = 22{,}000{,}000\ \text{ppb (calc and corr)}$

13.71 a) $523\ \text{mL air} \times \dfrac{3.0\ \text{g } CO_2}{10^9\ \text{mL air}} = 1.569 \times 10^{-6}\ \text{(calc)} = 1.6 \times 10^{-6}\ \text{g } CO_2\ \text{(corr)}$

b) $523\ \text{mL air} \times \dfrac{6.0\ \text{g } CO_2}{10^6\ \text{mL air}} = 3.138 \times 10^{-3}\ \text{(calc)} = 3.1 \times 10^{-3}\ \text{g } CO_2\ \text{(corr)}$

c) $523\ \text{mL air} \times \dfrac{2.5\ \text{g } CO_2}{10^2\ \text{mL air}} = 13.075\ \text{(calc)} = 13\ \text{g } CO_2\ \text{(corr)}$

d) $523\ \text{mL air} \times \dfrac{5.2\ \text{g } CO_2}{100\ \text{mL air}} = 27.196\ \text{(calc)} = 27\ \text{g } CO_2\ \text{(corr)}$

MOLARITY (SEC. 13.8)

13.73 $M = \dfrac{\text{moles solute}}{\text{L solution}}$;   molar mass HCl = 36.46 g/mole

a) $M = \dfrac{1.2 \text{ moles HCl}}{1.27 \text{ L solution}} = 0.944881889 \text{ (calc)} = 0.94 \text{ M (corr)}$

b) $775 \text{ mL} \times \dfrac{10^{-3} \text{ L}}{1 \text{ mL}} = 0.775 \text{ L}$

   $M = \dfrac{0.755 \text{ mole HCl}}{0.775 \text{ L solution}} = 0.974193548 \text{ (calc)} = 0.974 \text{ M (corr)}$

c) $1.2 \text{ g HCl} \times \dfrac{1 \text{ mole HCl}}{36.46 \text{ g HCl}} = 0.032912781 \text{ (calc)} = 0.033 \text{ mole HCl (corr)}$

   $M = \dfrac{0.033 \text{ mole HCl}}{0.025 \text{ L solution}} = 1.32 \text{ (calc)} = 1.3 \text{ M (corr)}$

d) $7.5 \text{ g HCl} \times \dfrac{1 \text{ mole HCl}}{36.46 \text{ g HCl}} = 0.205704882 \text{ (calc)} = 0.21 \text{ mole HCl (corr)}$

   $825 \text{ mL} \times \dfrac{10^{-3} \text{ L}}{1 \text{ mL}} = 0.825 \text{ L}$

   $M = \dfrac{0.21 \text{ mole HCl}}{0.825 \text{ L solution}} = 0.254545454 \text{ (calc)} = 0.25 \text{ M (corr)}$

13.75   Volume × molarity = moles

a) $1.25 \text{ L} \times \dfrac{0.500 \text{ mole KOH}}{\text{L solution}} = 0.625 \text{ mole KOH (calc and corr)}$

b) $1.25 \text{ mL} \times \dfrac{10^{-3} \text{ L}}{1 \text{ mL}} = 1.25 \times 10^{-3} \text{ L}$

   $1.25 \times 10^{-3} \text{ L} \times \dfrac{0.500 \text{ mole KOH}}{\text{L solution}} = 6.25 \times 10^{-4} \text{ mole KOH (calc and corr)}$

c) $62.0 \text{ L} \times \dfrac{2.00 \text{ moles KOH}}{\text{L solution}} = 124 \text{ moles KOH (calc and corr)}$

d) $743 \text{ mL} \times \dfrac{10^{-3} \text{ L}}{1 \text{ mL}} = 0.743 \text{ L}$

   $0.743 \text{ L} \times \dfrac{12.0 \text{ moles KOH}}{\text{L solution}} = 8.916 \text{ (calc)} = 8.92 \text{ moles KOH (corr)}$

13.77 $\text{volume} = \dfrac{\text{moles}}{\text{M}}$ NaCl = 58.44 g/mole

a) $\dfrac{10.0 \text{ moles NaCl}}{0.500 \text{ M}} = 20 \text{ (calc)} = 20.0 \text{ L (corr)}$

   $20.0 \text{ L} \times \dfrac{1 \text{ mL}}{10^{-3} \text{ L}} = 20.0 \times 10^3 \text{ mL}$

b) $10.0 \text{ g NaCl} \times \dfrac{1 \text{ mole NaCl}}{58.44 \text{ g NaCl}} = 0.171115674 \text{ (calc)} = 0.171 \text{ mole NaCl (corr)}$

$\dfrac{0.171 \text{ mole NaCl}}{0.500 \text{ M}} = 0.342 \text{ L solution (calc and corr)}$

$0.342 \text{ L} \times \dfrac{1 \text{ mL}}{10^{-3} \text{ L}} = 342 \text{ mL}$

c) $\dfrac{2.75 \text{ moles NaCl}}{0.80 \text{ M}} = 3.4375 \text{ (calc)} = 3.4 \text{ L (corr)}$

$3.4 \text{ L} \times \dfrac{1 \text{ mL}}{10^{-3} \text{ L}} = 3400 \text{ mL}$

d) $25.0 \text{ g NaCl} \times \dfrac{1 \text{ mole NaCl}}{58.44 \text{ g NaCl}} = 0.427789185 \text{ (calc)} = 0.428 \text{ mole NaCl (corr)}$

$\dfrac{0.428 \text{ mole NaCl}}{2.75 \text{ M}} = 0.155636363 \text{ (calc)} = 0.156 \text{ L (corr)}$

$0.156 \text{ L} \times \dfrac{1 \text{ mL}}{10^{-3} \text{ L}} = 156 \text{ mL}$

13.79 $M = \dfrac{1 \text{ mole NaOH}}{\text{L solution}};$ molar mass NaOH $= 40.00 \text{ g/mole}$

	grams solute	moles solute	liters solution	molarity
	80.0	2.00	1.75	1.14
a)	13.7	0.343	0.833	0.412
b)	10.0	0.250	0.179	1.40
c)	104	2.60	1.30	2.00
d)	0.100	0.00250	0.0270	0.0926

a) $0.343 \text{ mole NaOH} \times \dfrac{40.00 \text{ g NaOH}}{1 \text{ mole NaOH}} = 13.72 \text{ (calc)} = 13.7 \text{ g NaOH (corr)}$

$\dfrac{0.343 \text{ mole NaOH}}{0.833 \text{ L solution}} = 0.411764705 \text{ (calc)} = 0.412 \text{ M (corr)}$

b) $10.0 \text{ g NaOH} \times \dfrac{1 \text{ mole NaOH}}{40.00 \text{ g NaOH}} = 0.25 \text{ (calc)} = 0.250 \text{ mole NaOH (corr)}$

$\dfrac{0.250 \text{ mole NaOH}}{1.40 \text{ M}} = 0.178571428 \text{ (calc)} = 0.179 \text{ L (corr)}$

c) $1.30 \text{ L solution} \times \dfrac{2.00 \text{ moles NaOH}}{\text{L solution}} = 2.6 \text{ (calc)} = 2.60 \text{ moles NaOH (corr)}$

$2.60 \text{ moles NaOH} \times \dfrac{40.00 \text{ g NaOH}}{1 \text{ mole NaOH}} = 104 \text{ g NaOH (calc and corr)}$

d) $0.100 \text{ g NaOH} \times \dfrac{1 \text{ mole NaOH}}{40.00 \text{ g NaOH}} = 2.5 \times 10^{-3} \text{ (calc)} = 0.00250 \text{ mole NaOH (corr)}$

$\dfrac{0.00250 \text{ mole NaOH}}{0.0270 \text{ L solution}} = 0.092592592 \text{ (calc)} = 0.0926 \text{ M (corr)}$

13.81 Assuming the volume of Diagram I is 1.0 L and each dot represents 1 mole,

Diagram I $= \dfrac{14 \text{ moles}}{1 \text{ L}} = 14 \text{ M (calc and corr)}$ Diagram II $= \dfrac{10 \text{ moles}}{1 \text{ L}} = 10. \text{ M (calc and corr)}$

Diagram III $= \dfrac{7 \text{ moles}}{1/2 \text{ L}} = 14 \text{ M (calc and corr)}$ Diagram IV $= \dfrac{7 \text{ moles}}{1/4 \text{ L}} = 28 \text{ M (calc and corr)}$

a) Diagram IV has the highest concentration.
b) The concentration is the same in Diagrams I and III.

13.83 $\dfrac{88.00 \text{ g CH}_4\text{O}}{100.0 \text{ g solution}} \times \dfrac{0.8274 \text{ g solution}}{1 \text{ mL solution}} \times \dfrac{1 \text{ mL}}{10^{-3} \text{ L}} \times \dfrac{1 \text{ mole CH}_4\text{O}}{32.05 \text{ g CH}_4\text{O}} = 22.7180031 \text{ (calc)}$

$= 22.72 \text{ M CH}_4\text{O (corr)}$

13.85 $\dfrac{2.019 \text{ moles NaBr}}{1 \text{ L solution}} \times \dfrac{102.89 \text{ g NaBr}}{1 \text{ mole NaBr}} \times \dfrac{10^{-3} \text{ L solution}}{1 \text{ mL solution}} \times \dfrac{1 \text{ mL solution}}{1.157 \text{ g solution}} \times 100$

$= 17.954616 \text{ (calc)} = 17.96\% \text{ (m/m) (corr)}$

13.87 $\dfrac{20.0 \text{ g HCl}}{100 \text{ mL solution}} \times \dfrac{1 \text{ mL solution}}{10^{-3} \text{ L solution}} \times \dfrac{1 \text{ mole HCl}}{36.46 \text{ g HCl}} = 5.485463522 \text{ (calc)} = 5.49 \text{ M HCl (corr)}$

MOLARITY AND CHEMICAL EQUATIONS (SEC. 13.9)

13.89 $\text{Pb(NO}_3)_2(\text{aq}) + 2 \text{ NaCl(aq)} \rightarrow \text{PbCl}_2(\text{s}) + 2 \text{ NaNO}_3(\text{aq})$

$0.500 \text{ L NaCl} \times \dfrac{4.00 \text{ moles NaCl}}{1 \text{ L NaCl}} \times \dfrac{1 \text{ mole Pb(NO}_3)_2}{2 \text{ moles NaCl}} \times \dfrac{1 \text{ L Pb(NO}_3)_2 \text{ solution}}{1.00 \text{ mole Pb(NO}_3)_2}$

$= 1 \text{ (calc)} = 1.00 \text{ L (corr) Pb(NO}_3)_2 \text{ solution}$

13.91 $2 \text{ HNO}_3(\text{aq}) + 3 \text{ H}_2\text{S(aq)} \rightarrow 2 \text{ NO(g)} + 3 \text{ S(s)} + 4 \text{ H}_2\text{O(l)}$

$30.0 \text{ mL} \times \dfrac{10^{-3} \text{ L}}{1 \text{ mL}} \times \dfrac{12.0 \text{ moles HNO}_3}{1 \text{ L}} \times \dfrac{3 \text{ moles S}}{2 \text{ moles HNO}_3} \times \dfrac{32.06 \text{ g S}}{1 \text{ mole S}}$

$= 17.3124 \text{ (calc)} = 17.3 \text{ g S (corr)}$

13.93 $\text{Ni(s)} + \text{H}_2\text{SO}_4(\text{aq}) \rightarrow \text{NiSO}_4(\text{aq}) + \text{H}_2(\text{g})$

$18.0 \text{ g Ni} \times \dfrac{1 \text{ mole Ni}}{58.69 \text{ g Ni}} \times \dfrac{1 \text{ mole H}_2\text{SO}_4}{1 \text{ mole Ni}} \times \dfrac{1 \text{ L H}_2\text{SO}_4 \text{ solution}}{0.50 \text{ mole H}_2\text{SO}_4} \times \dfrac{1 \text{ mL}}{10^{-3} \text{ L}}$

$= 613.3924007 \text{ (calc)} = 610 \text{ mL H}_2\text{SO}_4 \text{ solution (corr)}$

13.95 $HNO_3(aq) + NaOH(aq) \rightarrow NaNO_3(aq) + H_2O(l)$

moles HNO_3 = 23.7 mL NaOH solution $\times \dfrac{10^{-3}\ L}{1\ mL} \times \dfrac{0.100\ mole\ NaOH}{1\ L\ solution} \times \dfrac{1\ mole\ HNO_3}{1\ mole\ NaOH}$

$\qquad\qquad\qquad\qquad\qquad\qquad\qquad\qquad\qquad = 0.00237$ mole HNO_3 (calc and corr)

volume HNO_3 = 37.5 mL $\times \dfrac{10^{-3}\ L}{1\ mL} = 0.0375$ L HNO_3 solution

$M = \dfrac{0.00237\ mole\ HNO_3}{0.0375\ L\ solution} = 0.0632$ M HNO_3 (calc and corr)

13.97 $8\ HNO_3(aq) + 3\ Cu(s) \rightarrow 3\ Cu(NO_3)_2(aq) + 2\ NO(g) + 4\ H_2O(l)$

50.0 mL solution $\times \dfrac{10^{-3}\ L}{1\ mL} \times \dfrac{6.0\ moles\ HNO_3}{1\ L\ solution} \times \dfrac{2\ moles\ NO}{8\ moles\ HNO_3} \times \dfrac{22.41\ L\ NO}{1\ mole\ NO}$

$\qquad\qquad\qquad\qquad\qquad\qquad\qquad\qquad\qquad = 1.68075$ (calc) = 1.7 L NO (corr)

13.99 $CO_2(g) + Ca(OH)_2(aq) \rightarrow CaCO_3(s) + H_2O(l)$

2.00 L $CO_2 \times \dfrac{1\ mole\ CO_2}{22.41\ L\ CO_2} \times \dfrac{1\ mole\ Ca(OH)_2}{1\ mole\ CO_2} = 0.08924587238$ (calc)

$\qquad\qquad\qquad\qquad\qquad\qquad\qquad\qquad\qquad = 0.0892$ mole $Ca(OH)_2$ (corr)

$M = \dfrac{0.0892\ mole\ Ca(OH)_2}{1.75\ L\ solution} = 0.05097142857$ (calc) = 0.0510 M $Ca(OH)_2$ (corr)

DILUTION (SEC. 13.10)

13.101 $M_1V_1 = M_2V_2$ where $M_1 = 0.400$ M, $V_1 = 25.0$ L and

a) $V_2 = 40.0$ mL; $M_2 = 0.400$ M $\times \dfrac{25.0\ mL}{40.0\ mL} = 0.25$ (calc) = 0.250 M (corr)

b) $V_2 = 87.0$ mL; $M_2 = 0.400$ M $\times \dfrac{25.0\ mL}{87.0\ mL} = 0.114942528$ (calc) = 0.115 M (corr)

c) $V_2 = 225$ mL; $M_2 = 0.400$ M $\times \dfrac{25.0\ mL}{225\ mL} = 0.04444$ (calc) = 0.0444 M (corr)

d) $V_2 = 1.45$ L $\times \dfrac{1\ mL}{10^{-3}\ L} = 1450$ mL; $M_2 = 0.400$ M $\times \dfrac{25.0\ mL}{1450\ mL} = 0.006896552$ (calc)

$\qquad\qquad\qquad\qquad\qquad\qquad\qquad\qquad\qquad = 0.00690$ M (corr)

13.103 $M_2 = M_1 \times \dfrac{V_1}{V_2}$ where $M_1 = 0.500$ M, $V_1 = 1353$ mL and

a) $V_2 = 1223$ mL; $M_2 = 0.500$ M $\times \dfrac{1353\ mL}{1223\ mL} = 0.553147996$ (calc) = 0.553 M (corr)

b) $V_2 = 1.12$ L $\times \dfrac{1\ mL}{10^{-3}\ L} = 1120$ mL; $M_2 = 0.500$ M $\times \dfrac{1353\ mL}{1120\ mL} = 0.604017857$ (calc)

$\qquad\qquad\qquad\qquad\qquad\qquad\qquad\qquad\qquad = 0.604$ M (corr)

c) $V_2 = 853$ mL; $M_2 = 0.500$ M $\times \dfrac{1353 \text{ mL}}{853 \text{ mL}} = 0.793083235$ (calc) $= 0.793$ M (corr)

d) $V_2 = 302.5$ mL; $M_2 = 0.500$ M $\times \dfrac{1353 \text{ mL}}{302.5 \text{ mL}} = 2.2363636$ (calc) $= 2.24$ M (corr)

13.105 Dilutions: $M_{\text{initial}} \times V_{\text{initial}} = M_{\text{final}} \times V_{\text{final}}$

	initial volume	initial molarity	final volume	final molarity
	20.0 mL	1.20	30.0 mL	0.800
a)	1.26 L	2.10	1.20 L	2.20
b)	32.5 mL	2.49	1.12 L	0.0723
c)	1.12 L	0.725	1.30 L	0.625
d)	43.2 mL	11.2	83.2 mL	5.82

a) $V_{\text{initial}} = \dfrac{2.20\,M \times 1.20 \text{ L}}{2.10\,M} = 1.257142857$ (calc) $= 1.26$ L (corr)

b) 32.5 mL $\times \dfrac{10^{-3} \text{ L}}{1 \text{ mL}} = 0.0325$ L

  $M_{\text{initial}} = \dfrac{0.0723 \text{ M} \times 1.12 \text{ L}}{0.0325 \text{ L}} = 2.491569231$ (calc) $= 2.49$ M (corr)

c) $V_{\text{final}} = \dfrac{0.725\,M \times 1.12 \text{ L}}{0.625\,M} = 1.2992$ (calc) $= 1.30$ L (corr)

d) $M_{\text{final}} = \dfrac{11.2 \text{ M} \times 43.2 \text{ mL}}{83.2 \text{ mL}} = 5.815384615$ (calc) $= 5.82$ M (corr)

13.107 Diagram III

13.109 a) $V_1 = 20.0$ mL; $M_1 = 2.00$ M; $V_2 = 20.0$ mL $\times \dfrac{2.00 \text{ M}}{0.100 \text{ M}} = 40\overline{0}$ mL (calc and corr)

  $V_{\text{H}_2\text{O}} = 40\overline{0}$ mL $- 20.0$ mL $= 38\overline{0}$ mL H_2O

b) $V_1 = 20.0$ mL; $M_1 = 0.250$ M; $V_2 = 20.0$ mL $\times \dfrac{0.250 \text{ M}}{0.100 \text{ M}} = 50$ mL (calc) $= 50.0$ mL (corr)

  $V_{\text{H}_2\text{O}} = 50.0 - 20.0 = 30.0$ mL H_2O

c) $V_1 = 358$ mL; $M_1 = 0.950$ M; $V_2 = 358$ mL $\times \dfrac{0.950 \text{ M}}{0.100 \text{ M}} = 3401$ mL (calc) $= 340\overline{0}$ mL (corr)

  $V_{\text{H}_2\text{O}} = 34\overline{0}0 - 358 = 3042$ (calc) $= 3040$ mL H_2O (corr)

d) $V_1 = 2.3$ L; $M_1 = 6.00$ M; $V_2 = 2.3$ L $\times \dfrac{6.00 \text{ M}}{0.100 \text{ M}} = 138$ L (calc) $= 140$ L (corr)

  $V_{\text{H}_2\text{O}} = 140$ L $- 2.3$ L $= 137.7$ (calc) $= 140$ (corr) $\times \dfrac{1 \text{ mL}}{10^{-3} \text{ L}} = 1.4 \times 10^5$ mL H_2O

13.111 a) $6.0 \text{ M} \times \dfrac{25.0 \text{ mL}}{45.0 \text{ mL}} = 3.3333333 \text{ (calc)} = 3.3 \text{ M (corr)}$

b) $3.0 \text{ M} \times \dfrac{100.0 \text{ mL}}{120.0 \text{ mL}} = 2.5 \text{ M (calc and corr)}$

c) $10.0 \text{ M} \times \dfrac{155 \text{ mL}}{175 \text{ mL}} = 8.8571429 \text{ (calc)} = 8.86 \text{ M (corr)}$

d) $0.100 \text{ M} \times \dfrac{2.00 \text{ mL}}{22.0 \text{ mL}} = 0.009090909 \text{ (calc)} = 0.00909 \text{ M (corr)}$

13.113 a) $V_2 = 275 \text{ mL} + 3.254 \text{ L} \times \dfrac{1 \text{ mL}}{10^{-3} \text{ L}} = 3529 \text{ mL}$

$M_2 = 6.00 \text{ M} \times \dfrac{275 \text{ mL}}{3529 \text{ mL}} = 0.4675545 \text{ (calc)} = 0.468 \text{ M (corr)}$

b) simple method: 6.00 M. Both solutions are 6.00 M, hence the mixture is 6.00 M. See part (c) for a model to work the problem rigorously.

c) moles NaOH (solution 1) $= \dfrac{6.00 \text{ moles NaOH}}{1 \text{ L solution}} \times 275 \text{ mL} \times \dfrac{10^{-3} \text{ L}}{1 \text{ mL}}$
$= 1.65 \text{ moles NaOH (calc and corr)}$

moles NaOH (solution 2) $= \dfrac{2.00 \text{ moles NaOH}}{1 \text{ L solution}} \times 125 \text{ mL} \times \dfrac{10^{-3} \text{ L}}{1 \text{ mL}}$
$= 0.250 \text{ mole NaOH (calc and corr)}$

total moles $= 1.65 + 0.250 = 1.90 \text{ moles NaOH}$

total volume $= 275 + 125 = 400 \text{ mL} \times \dfrac{1 \text{ L}}{10^{-3} \text{ mL}} = 0.400 \text{ L solution}$

$M_{\text{final}} = \dfrac{1.90 \text{ moles NaOH}}{0.400 \text{ L solution}} = 4.75 \text{ M NaOH (calc and corr)}$

d) moles NaOH (solution 1) $= \dfrac{6.00 \text{ moles NaOH}}{1 \text{ L solution}} \times 275 \text{ mL} \times \dfrac{10^{-3} \text{ L}}{1 \text{ mL}}$
$= 1.65 \text{ moles (calc and corr)}$

moles NaOH (solution 2) $= \dfrac{5.80 \text{ moles NaOH}}{1 \text{ L solution}} \times 27 \text{ mL} \times \dfrac{10^{-3} \text{ L}}{1 \text{ mL}}$
$= 0.1566 \text{ (calc)} = 0.16 \text{ mole NaOH (corr)}$

total moles $= 1.65 + 0.16 = 1.807 \text{ (calc)} = 1.81 \text{ moles NaOH (corr)}$

total volume (L) $= 275 \text{ mL} + 27 \text{ mL} = 302 \text{ mL} \times \dfrac{10^{-3} \text{ L}}{1 \text{ mL}} = 0.302 \text{ L solution (calc and corr)}$

$M_{\text{final}} = \dfrac{1.81 \text{ moles NaOH}}{0.302 \text{ L solution}} = 5.9933775 \text{ (calc)} = 5.99 \text{ M NaOH (corr)}$

MOLALITY (SEC. 13.11)

13.115 $\text{molality} = \dfrac{\text{moles HNO}_3}{\text{kg H}_2\text{O}}$ molar mass $\text{HNO}_3 = 63.02$ g/mole

a) $16.5 \text{ g HNO}_3 \times \dfrac{1 \text{ mole HNO}_3}{63.02 \text{ g HNO}_3} = 0.261821643 \text{ (calc)} = 0.262 \text{ mole HNO}_3 \text{ (corr)}$

$\dfrac{0.262 \text{ mole HNO}_3}{1.35 \text{ kg H}_2\text{O}} = 0.194074074 \text{ (calc)} = 0.194 \text{ m (corr)}$

b) $455 \text{ g} \times \dfrac{1 \text{ kg}}{10^3 \text{ g}} = 0.455 \text{ kg}$

$\dfrac{3.15 \text{ moles HNO}_3}{0.455 \text{ kg H}_2\text{O}} = 6.923076923 \text{ (calc)} = 6.92 \text{ m (corr)}$

c) $13.0 \text{ g} \times \dfrac{1 \text{ kg}}{10^3 \text{ g}} = 0.0130 \text{ kg}$

$0.0756 \text{ g HNO}_3 \times \dfrac{1 \text{ mole HNO}_3}{63.02 \text{ g HNO}_3} = 0.001199619169 \text{ (calc)} = 0.00120 \text{ mole HNO}_3 \text{ (corr)}$

$\dfrac{0.00120 \text{ mole HNO}_3}{0.0130 \text{ kg H}_2\text{O}} = 0.092307692 \text{ (calc)} = 0.0923 \text{ m HNO}_3 \text{ (corr)}$

d) $\dfrac{0.0022 \text{ mole HNO}_3}{0.355 \text{ kg H}_2\text{O}} = 6.197183099 \times 10^{-3} \text{ (calc)} = 6.2 \times 10^{-3} \text{ m (corr)}$

13.117 a) $264 \text{ g H}_2\text{O} \times \dfrac{1 \text{ kg H}_2\text{O}}{10^3 \text{ g H}_2\text{O}} \times \dfrac{0.750 \text{ mole KCl}}{\text{kg H}_2\text{O}} \times \dfrac{74.55 \text{ g KCl}}{1 \text{ mole KCl}} = 14.7609 \text{ (calc)} = 14.8 \text{ g KCl (corr)}$

b) $0.100 \text{ kg H}_2\text{O} \times \dfrac{0.333 \text{ mole NH}_4\text{NO}_3}{\text{kg H}_2\text{O}} \times \dfrac{80.06 \text{ g NH}_4\text{NO}_3}{1 \text{ mole NH}_4\text{NO}_3}$
$= 2.665998 \text{ (calc)} = 2.67 \text{ g NH}_4\text{NO}_3 \text{ (corr)}$

c) $1.20 \text{ kg H}_2\text{O} \times \dfrac{1.22 \text{ mole K}_3\text{PO}_4}{\text{kg H}_2\text{O}} \times \dfrac{212.27 \text{ g K}_3\text{PO}_4}{1 \text{ mole K}_3\text{PO}_4} = 310.76328 \text{ (calc)} = 311 \text{ g K}_3\text{PO}_4 \text{ (corr)}$

d) $23 \text{ g H}_2\text{O} \times \dfrac{1 \text{ kg H}_2\text{O}}{10^3 \text{ g H}_2\text{O}} \times \dfrac{6.00 \text{ mole NaOH}}{\text{kg H}_2\text{O}} \times \dfrac{40.00 \text{ g NaOH}}{1 \text{ mole NaOH}} = 5.52 \text{ g NaOH (calc and corr)}$

13.119 $m = \dfrac{1 \text{ mole NaOH}}{\text{kg H}_2\text{O}}$;   molar mass NaOH $= 40.00$ g/mole

	moles solute	grams solute	kilograms solvent	molality
	2.00	80.0	1.20	1.67
a)	4.27	171	1.33	3.21
b)	1.08	43.0	0.872	1.24
c)	0.334	13.4	2.00	0.167
d)	0.100	4.00	0.704	0.142

a) $1.33 \text{ kg H}_2\text{O} \times \dfrac{3.21 \text{ moles NaOH}}{\text{kg H}_2\text{O}} = 4.2693 \text{ (calc)} = 4.27 \text{ moles NaOH (calc)}$

$4.27 \text{ moles NaOH} \times \dfrac{40.00 \text{ g NaOH}}{1 \text{ mole NaOH}} = 170.8 \text{ (calc)} = 171 \text{ g NaOH (corr)}$

b) $43.0 \text{ g NaOH} \times \dfrac{1 \text{ mole NaOH}}{40.00 \text{ g NaOH}} = 1.075 \text{ (calc)} = 1.08 \text{ mole NaOH (corr)}$

$\dfrac{1.08 \text{ moles NaOH}}{0.872 \text{ kg H}_2\text{O}} = 1.23853211 \text{ (calc)} = 1.24 \text{ m (corr)}$

c) $0.334 \text{ mole NaOH} \times \dfrac{40.00 \text{ g NaOH}}{1 \text{ mole NaOH}} = 13.36 \text{ (calc)} = 13.4 \text{ g NaOH (corr)}$

$\dfrac{0.334 \text{ mole NaOH}}{2.00 \text{ kg H}_2\text{O}} = 0.167 \text{ m (calc and corr)}$

d) $0.100 \text{ moles NaOH} \times \dfrac{40.00 \text{ g NaOH}}{1 \text{ mole NaOH}} = 4 \text{ (calc)} = 4.00 \text{ g NaOH (corr)}$

$0.100 \text{ moles NaOH} \times \dfrac{\text{kg H}_2\text{O}}{0.142 \text{ mole NaOH}} = 0.704225352 \text{ (calc)} = 0.704 \text{ kg H}_2\text{O (corr)}$

13.121 $\text{Mass of C}_6\text{H}_{12} = 75.0 \text{ mL} \times \dfrac{0.779 \text{ g C}_6\text{H}_{12}}{1 \text{ mL C}_6\text{H}_{12}} = 58.425 \text{ (calc)} = 58.4 \text{ g C}_6\text{H}_{12} \text{ (corr)}$

$\text{Mass of C}_6\text{H}_{14} = 175.0 \text{ mL} \times \dfrac{0.659 \text{ g C}_6\text{H}_{14}}{1 \text{ mL C}_6\text{H}_{14}} = 115.325 \text{ (calc)} = 115 \text{ g C}_6\text{H}_{14} \text{ (corr)}$

Therefore C_6H_{12} is the solute, and C_6H_{14} is the solvent.

$\text{Moles of solute} = 58.4 \text{ g C}_6\text{H}_{12} \times \dfrac{1 \text{ mole C}_6\text{H}_{12}}{84.18 \text{ g C}_6\text{H}_{12}} = 0.693751484 \text{ (calc)} = 0.694 \text{ mole C}_6\text{H}_{12} \text{ (corr)}$

$\text{kg of solvent} = 115 \text{ g C}_6\text{H}_{14} \times \dfrac{1 \text{ kg}}{10^3 \text{ g}} = \text{ (calc)} = 0.115 \text{ kg C}_6\text{H}_{14} \text{ (corr)}$

$\text{Molality} = \dfrac{\text{moles solute}}{\text{kg solvent}} = \dfrac{0.694 \text{ mole C}_6\text{H}_{12}}{0.115 \text{ kg C}_6\text{H}_{14}} = 6.034782609 \text{ (calc)} = 6.03 \text{ m C}_6\text{H}_{12} \text{ (corr)}$

13.123 Basis: 1 L solution
Mass of solute:

$1 \text{ L solution} \times \dfrac{0.568 \text{ mole H}_2\text{C}_2\text{O}_4}{1 \text{ L solution}} \times \dfrac{90.04 \text{ g H}_2\text{C}_2\text{O}_4}{1 \text{ mole H}_2\text{C}_2\text{O}_4} = 51.14272 \text{ (calc)} = 51.1 \text{ g H}_2\text{C}_2\text{O}_4 \text{ (corr)}$

Mass of solution:

$1 \text{ L solution} \times \dfrac{1 \text{ mL solution}}{10^{-3} \text{ L solution}} \times \dfrac{1.022 \text{ g solution}}{1 \text{ mL solution}} = 1022 \text{ g solution (calc and corr)}$

Mass of solvent: $(1022 - 51.1) \text{ g} = 970.9 \text{ (calc)} = 971 \text{ g (corr)}$

Molality: $\dfrac{51.1 \text{ g H}_2\text{C}_2\text{O}_4}{971 \text{ g H}_2\text{O}} \times \dfrac{1 \text{ mole H}_2\text{C}_2\text{O}_4}{90.04 \text{ g H}_2\text{C}_2\text{O}_4} \times \dfrac{10^3 \text{ g H}_2\text{O}}{1 \text{ kg H}_2\text{O}} = 0.5844753 \text{ (calc)}$
$$= 0.584 \text{ m H}_2\text{C}_2\text{O}_4 \text{ (corr)}$$

13.125   Basis: 1 kg solvent = 1000 g solvent
Mass of solute:

$$1 \text{ kg solvent} \times \frac{0.796 \text{ mole HC}_2\text{H}_3\text{O}_2}{1 \text{ kg solvent}} \times \frac{60.06 \text{ g HC}_2\text{H}_3\text{O}_2}{1 \text{ mole HC}_2\text{H}_3\text{O}_2} = 47.80776 \text{ (calc)}$$

$$= 47.8 \text{ g HC}_2\text{H}_3\text{O}_2 \text{ (corr)}$$

Mass of solution: (1000 + 47.8) g = 1047.8 g (calc and corr)

Volume of solution: $1047.8 \text{ g solution} \times \dfrac{1 \text{ mL solution}}{1.004 \text{ g solution}} \times \dfrac{10^{-3} \text{ L solution}}{1 \text{ mL solution}}$

$$= 1.043625498 \text{ (calc)} = 1.044 \text{ L solution (corr)}$$

Molarity: $\dfrac{47.8 \text{ g HC}_2\text{H}_3\text{O}_2}{1.044 \text{ L solution}} \times \dfrac{1 \text{ mole HC}_2\text{H}_3\text{O}_2}{60.06 \text{ g HC}_2\text{H}_3\text{O}_2} = 0.7623283485 \text{ (calc)} = 0.762 \text{ M HC}_2\text{H}_3\text{O}_2 \text{ (corr)}$

13.127   Basis: 1 L of solution = 10^3 mL solution
Mass of solvent:

$$1 \text{ L solution} \times \frac{11.8 \text{ moles HCl}}{1 \text{ L solution}} \times \frac{1 \text{ kg solvent}}{15.5 \text{ moles HCl}} \times \frac{10^3 \text{ g}}{1 \text{ kg}} = 761.2903226 \text{ (calc)}$$

$$= 761 \text{ g solvent (corr)}$$

Mass of solute:

$$1 \text{ L solution} \times \frac{11.8 \text{ moles HCl}}{1 \text{ L solution}} \times \frac{36.46 \text{ g HCl}}{1 \text{ mole HCl}} = 430.228 \text{ (calc)} = 430. \text{ g HCl (corr)}$$

Mass of solution = 761 g solvent + 430. g solute = 1191 g solution (calc and corr)

density of solution = $\dfrac{\text{mass of solution}}{\text{volume of solution}} = \dfrac{1191 \text{ g solution}}{10^3 \text{ mL solution}} = 1.191 \text{ g/mL (calc and corr)}$

13.129   Mass of solvent = 100 − 22.0 g solute = 78.0 g solvent

$$\frac{22.0 \text{ g H}_8\text{C}_6\text{O}_7}{78.0 \text{ g solvent}} \times \frac{1 \text{ mole H}_8\text{C}_6\text{O}_7}{192.14 \text{ g H}_8\text{C}_6\text{O}_7} \times \frac{10^3 \text{ g solvent}}{1 \text{ kg solvent}} = 1.467946716 \text{ (calc)} = 1.47 \text{ m H}_8\text{C}_6\text{O}_7 \text{ (corr)}$$

Multi-Concept Problems

13.131   a) $(NH_4)_3PO_4$ b) $Ca(OH)_2$ c) $AgNO_3$ d) $CaS, Ca(NO_3)_2, Ca(C_2H_3O_2)_2$

13.133   Since the solution is 75% saturated, it contains

$0.75 \times 33.3 \text{ g CuSO}_4 = 24.975 \text{ (calc)} = 25 \text{ g CuSO}_4$ in 100 g H_2O.

The mass % = $\dfrac{25 \text{ g}}{(100 + 25) \text{ g solution}} \times 100 = 2\bar{0}\% \text{ (m/m) CuSO}_4 \text{ (calc and corr)}$

$400.0 \text{ g solution} \times \dfrac{2\bar{0} \text{ g CuSO}_4}{100 \text{ g solution}} = 8\bar{0} \text{ g CuSO}_4 \text{ (calc and corr)}$

13.135 a) $\%\dfrac{m}{v} = \dfrac{45.2 \text{ g AgNO}_3}{254 \text{ mL solution}} \times 100 = 17.7952756 \text{ (calc)} = 17.8\% \text{ (m/v) (corr)}$

b) $\dfrac{45.2 \text{ g AgNO}_3 \times \dfrac{1 \text{ mole AgNO}_3}{169.88 \text{ g AgNO}_3}}{254 \text{ mL solution} \times \dfrac{10^{-3} \text{ L}}{1 \text{ mL}}} = 1.0475203 \text{ (calc)} = 1.05 \text{ M AgNO}_3 \text{ (corr)}$

13.137 $425 \text{ mL solution} \times \dfrac{1.02 \text{ g solution}}{1 \text{ mL solution}} \times \dfrac{1.55 \text{ g Na}_2\text{SO}_4}{100 \text{ g solution}} = 6.71925 \text{ (calc)} = 6.72 \text{ g Na}_2\text{SO}_4 \text{ (corr)}$

13.139 Grams of solvent = grams of solution − grams of solute
Grams of solution:

$1375 \text{ mL solution} \times \dfrac{1.161 \text{ g solution}}{1 \text{ mL solution}} = 1596.375 \text{ (calc)} = 1596 \text{ g solution (corr)}$

Grams of solute:

$1.375 \text{ L solution} \times \dfrac{3.000 \text{ moles NaNO}_3}{1 \text{ L solution}} \times \dfrac{85.00 \text{ g NaNO}_3}{1 \text{ mole NaNO}_3}$
$$= 350.625 \text{ (calc)} = 350.6 \text{ g NaNO}_3 \text{ (corr)}$$

Grams of solvent: $(1596 - 350.6) \text{ g} = 1245.4 \text{ (calc)} = 1245 \text{ g H}_2\text{O (corr)}$

13.141 $3.74 \text{ ppm (m/m)} = \dfrac{3.74 \text{ mg solute}}{10^6 \text{ mg solution}}$

$\dfrac{3.74 \text{ mg solute}}{10^6 \text{ mg solution}} \times \dfrac{1 \text{ mg solution}}{10^{-3} \text{ g solution}} \times \dfrac{10^3 \text{ g solution}}{1 \text{ kg solution}} = 3.74 \dfrac{\text{mg solute}}{\text{kg solution}} \text{ (calc and corr)}$

13.143 $1.00 \text{ g NaCl} \times \dfrac{1 \text{ mole NaCl}}{58.44 \text{ g NaCl}} = 0.01711156742 \text{ (calc)} = 0.0171 \text{ mole NaCl (corr)}$

$M = \dfrac{0.0171 \text{ mole NaCl}}{0.01000 \text{ L solution}} = 1.71 \text{ M NaCl (calc and corr)}$

$M_2 = 1.71 \text{ M} \times \dfrac{1.00 \text{ mL}}{10.00 \text{ mL}} = 0.171 \text{ M NaCl (calc and corr)}$

13.145 Molarity of a 38.0% (m/m) HCl solution: Basis: $1\overline{000}$ g of solution

$1\overline{000} \text{ g solution} \times \dfrac{38.0 \text{ g HCl}}{100 \text{ g solution}} \times \dfrac{1 \text{ mole HCl}}{36.46 \text{ g HCl}} = 10.422381 \text{ (calc)} = 10.4 \text{ moles HCl (corr)}$

$1\overline{000} \text{ g solution} \times \dfrac{1 \text{ mL solution}}{1.19 \text{ g solution}} \times \dfrac{10^{-3} \text{ L solution}}{1 \text{ mL solution}} = 0.84033613 \text{ (calc)} = 0.840 \text{ L solution (corr)}$

$M_1 = \dfrac{10.4 \text{ moles HCl}}{0.840 \text{ L solution}} = 12.380952 \text{ (calc)} = 12.4 \text{ M HCl (corr)}$

Dilution: $V_1 = 1\overline{000} \text{ mL} \times \dfrac{0.100 \text{ M}}{12.4 \text{ M}} = 8.0645161 \text{ (calc)} = 8.06 \text{ mL solution (corr)}$

13.147   Moles of NaCl in original solution:

$$1.50 \text{ kg H}_2\text{O} \times \frac{1.23 \text{ moles NaCl}}{1 \text{ kg H}_2\text{O}} = 1.845 \text{ (calc)} = 1.84 \text{ moles NaCl (corr)}$$

Mass of H$_2$O in final solution:

$$1.84 \text{ moles NaCl} \times \frac{1 \text{ kg H}_2\text{O}}{1.00 \text{ mole NaCl}} = 1.84 \text{ kg H}_2\text{O (calc and corr)}$$

$$(1.84 - 1.50) \text{ kg H}_2\text{O} = 0.34 \text{ kg H}_2\text{O} \times \frac{10^3 \text{ g H}_2\text{O}}{1 \text{ kg H}_2\text{O}} = 340 \text{ g H}_2\text{O (calc and corr)}$$

13.149   Mass of solution = mass of solute + mass of solvent
Mass of solute = 52.0 g
Mass of solvent:

$$52.0 \text{ g H}_3\text{PO}_4 \times \frac{1 \text{ mole H}_3\text{PO}_4}{98.00 \text{ g H}_3\text{PO}_4} \times \frac{1 \text{ kg solvent}}{2.16 \text{ moles H}_3\text{PO}_4} \times \frac{10^3 \text{ g solvent}}{1 \text{ kg solvent}}$$
$$= 245.6538171 \text{ (calc)} = 246 \text{ g solvent (corr)}$$

Mass of solution = (52.0 + 246) g = 298 g solution (calc and corr)
Volume of solution:

$$298 \text{ g solution} \times \frac{1 \text{ mL solution}}{1.12 \text{ g solution}} = 266.0714286 \text{ (calc)} = 266 \text{ mL solution (corr)}$$

13.151   a) $11.3 \text{ mL CH}_3\text{OH} \times \dfrac{0.793 \text{ g CH}_3\text{OH}}{1 \text{ mL CH}_3\text{OH}} = 8.9609 \text{ (calc)} = 8.96 \text{ g CH}_3\text{OH (corr)}$

$$\%(\text{m/v}) = \frac{8.96 \text{ g CH}_3\text{OH}}{75.0 \text{ mL solution}} \times 100 = 11.946666 \text{ (calc)} = 11.9\% \text{ (m/v) CH}_3\text{OH (corr)}$$

b) $75.0 \text{ mL solution} \times \dfrac{0.980 \text{ g solution}}{1 \text{ mL solution}} = 73.5 \text{ g solution (calc and corr)}$

$$\%(\text{m/m}) = \frac{8.96 \text{ g CH}_3\text{OH}}{73.5 \text{ g solution}} \times 100 = 12.19047619 \text{ (calc)} = 12.2\% \text{ (m/m) CH}_3\text{OH (corr)}$$

c) $\%(\text{v/v}) = \dfrac{11.3 \text{ mL CH}_3\text{OH}}{75.0 \text{ mL solution}} \times 100 = 15.066666 \text{ (calc)} = 15.1\% \text{ (v/v) CH}_3\text{OH (corr)}$

13.153   Basis: 1 mole KCl

Volume of solution: $1 \text{ mole KCl} \times \dfrac{1 \text{ L solution}}{0.271 \text{ mole KCl}} \times \dfrac{1 \text{ mL solution}}{10^{-3} \text{ L solution}}$
$$= 3690.0369 \text{ (calc)} = 3690 \text{ mL solution (corr)}$$

Mass of solution = mass of solute + mass of solvent

Mass of solute: $1 \text{ mole KCl} \times \dfrac{74.55 \text{ g KCl}}{1 \text{ mole KCl}} = 74.55 \text{ g KCl (calc and corr)}$

Mass of solvent: $1 \text{ mole KCl} \times \dfrac{1 \text{ kg solvent}}{0.273 \text{ mole KCl}} \times \dfrac{10^3 \text{ g solvent}}{1 \text{ kg solvent}} = 3663.003663 \text{ (calc)}$
$$= 3660 \text{ g solvent (corr)}$$

Mass of solution: $(3660 + 74.55)$ g $= 3734.55$ (calc) $= 3730$ g (corr)

$$\text{density} = \frac{\text{mass solution}}{\text{volume solution}} = \frac{3730 \text{ g solution}}{3690 \text{ mL solution}} = 1.0108401 \text{ (calc)} = 1.01 \text{ g/mL (corr)}$$

Answers to Multiple-Choice Practice Test

MC 13.1 a	**MC 13.2** b	**MC 13.3** d	**MC 13.4** e	**MC 13.5** b	**MC 13.6** c
MC 13.7 b	**MC 13.8** b	**MC 13.9** d	**MC 13.10** d	**MC 13.11** b	**MC 13.12** d
***MC 13.13** d	**MC 13.14** b	**MC 13.15** b	**MC 13.16** d	**MC 13.17** b	**MC 13.18** c
MC 13.19 d	**MC 13.20** d				

*MC 13.13 in text has the answer c, correct answer is d.

CHAPTER FOURTEEN
Acids, Bases, and Salts

Practice Problems by Topic

ACID–BASE DEFINITIONS (SECS. 14.1 AND 14.2)

14.1 a) The species responsible for the properties of acidic solutions is the hydrogen ion, $H^+(aq)$.
  b) The term used to describe formation of ions, in aqueous solution, from an ionic compound is *dissociation*.

14.3 a) Arrhenius acid   b) Arrhenius acid

14.5 a) ionization, HNO_3, molecular compound   b) dissociation, NaOH, ionic compound
  c) dissociation, RbOH, ionic compound   d) ionization, HCN, molecular compound

14.7 a) $HBr \rightarrow H^+ + Br^-$ b) $HClO_2 \rightarrow H^+ + ClO_2^-$
  c) $LiOH \rightarrow Li^+ + OH^-$ d) $Ba(OH)_2 \rightarrow Ba^{2+} + 2\, OH^-$

14.9 Brønsted–Lowry = BL
  a) $HBr + H_2O \rightarrow H_3O^+ + Br^-$ b) $H_2O + N_3^- \rightarrow HN_3 + OH^-$
  BL acid + BL base   BL acid + BL base
  c) $H_2O + H_2S \rightarrow H_3O^+ + HS^-$ d) $HS^- + H_2O \rightarrow H_3O^+ + S^{2-}$
  BL base + BL acid   BL acid + BL base

14.11 a) HOCl behaves as a Bronsted-Lowry acid:
  $HOCl(aq) + H_2O(l) \rightarrow OCl^-(aq) + H_3O^+(aq)$
  b) NH_3 behaves as a Bronsted-Lowry base:
  $NH_3(aq) + H_2O(l) \rightarrow NH_4^+(aq) + OH^-(aq)$
  c) $H_2PO_4^-$ behaves as a Bronsted-Lowry acid:
  $H_2PO_4^-(aq) + H_2O(l) \rightarrow HPO_4^{2-}(aq) + H_3O^+(aq)$
  d) CO_3^{2-} behaves as a Bronsted-Lowry base:
  $CO_3^{2-}(aq) + H_2O(l) \rightarrow HCO_3^-(aq) + OH^-(aq)$

14.13 a) $HOCl + NH_3 \rightarrow NH_4^+ + ClO^-$ b) $H_2CO_3 + H_2O \rightarrow HCO_3^- + H_3O^+$
  c) $H_2O + H_2O \rightarrow H_3O^+ + OH^-$ d) $HC_2O_4^- + H_2O \rightarrow C_2O_4^{2-} + H_3O^+$

CONJUGATE ACIDS AND BASES (SEC. 14.3)

14.15 a) The conjugate acid of HS^- is H_2S.   b) The conjugate base of HNO_2 is NO_2^-.
  c) The conjugate acid of PH_3 is PH_4^+.   d) The conjugate base of $H_2PO_4^-$ is HPO_4^{2-}.

14.17 Conjugate acid–base pairs
  a) $H_2C_2O_4$ and $HC_2O_4^-$; HClO and ClO^- b) HSO_4^- and SO_4^{2-}; H_3O^+ and H_2O
  c) $H_2PO_4^-$ and HPO_4^{2-}; NH_4^+ and NH_3 d) H_2CO_3 and HCO_3^-; H_2O and OH^-

14.19 a) NH_4^+/NH_3 is a Bronsted-Lowry conjugate acid-base pair.
b) HN_3/N_3^- is a Bronsted-Lowry conjugate acid-base pair.
c) H_3PO_4/PO_4^{3-} is not a Bronsted-Lowry conjugate acid-base pair.
d) SO_4^{2-}/SO_3^{2-} is not a Bronsted-Lowry conjugate acid-base pair.

14.21 a) (1) $HS^- + H_3O^+ \rightarrow H_2S + H_2O$ (2) $HS^- + OH^- \rightarrow H_2O + S^{2-}$
b) (1) $HPO_4^{2-} + H_3O^+ \rightarrow H_2PO_4^- + H_2O$ (2) $HPO_4^{2-} + OH^- \rightarrow H_2O + PO_4^{3-}$
c) (1) $HCO_3^- + H_3O^+ \rightarrow H_2CO_3 + H_2O$ (2) $HCO_3^- + OH^- \rightarrow H_2O + CO_3^{2-}$
d) (1) $H_2PO_3^- + H_3O^+ \rightarrow H_3PO_3 + H_2O$ (2) $H_2PO_3^- + OH^- \rightarrow H_2O + HPO_3^{2-}$

MONO-, DI-, AND TRIPROTIC ACIDS (SEC. 14.4)

14.23 a) $HClO_3$, monoprotic   b) $HC_3H_5O_4$, monoprotic
c) $HC_2H_3O_2$, monoprotic   d) $H_2C_4H_2O_4$, diprotic

14.25 a) $HClO_3$, monoprotic, 1 acidic H   b) $HC_3H_5O_4$, monoprotic, 1 acidic H
c) $HC_2H_3O_2$, monoprotic, 1 acidic H   d) $H_2C_4H_2O_4$, diprotic, 2 acidic H's

14.27 a) $HClO_3$, 0 nonacidic H's   b) $HC_3H_5O_4$, 5 nonacidic H's
c) $HC_2H_3O_2$, 3 nonacidic H's   d) $H_2C_4H_2O_4$, 2 nonacidic H's

14.29 a) i) $H_2C_2O_4 + H_2O \rightarrow HC_2O_4^- + H_3O^+$ ii) $HC_2O_4^- + H_2O \rightarrow C_2O_4^{2-} + H_3O^+$
b) i) $H_2C_3H_2O_4 + H_2O \rightarrow HC_3H_2O_4^- + H_3O^+$ ii) $HC_3H_2O_4^- + H_2O \rightarrow C_3H_2O_4^{2-} + H_3O^+$

14.31 The chemical formula for lactic acid, $H_6C_3O_3$, is written as $HC_3H_5O_3$, to differentiate between the 5 nonacidic H's and the 1 acidic H in the acid.

14.33 Pyruvic acid contains one acidic H, attached to the oxygen atom, and 3 nonacidic H's, attached to carbon.

STRENGTH OF ACIDS AND BASES (SEC. 14.5)

14.35 a) $HClO_3$ is a strong acid.   b) $HC_3H_5O_4$ is a weak acid.
c) $HC_2H_3O_2$ is a weak acid.   d) $H_2C_4H_2O_4$ is a weak acid.

14.37 a) H_2SO_4 is stronger than H_2SO_3.   b) $HClO_4$ is stronger than HCN.
c) HF is weaker than HBr.   d) HNO_2 is weaker than $HClO_3$.

14.39 a) strong acid, strong base   b) strong acid, strong base
c) weak acid, strong base   d) weak acid, strong base

14.41 a) HNO_3 is a strong acid, ionizes nearly 100% in water.
b) H_3PO_4 is a weak acid, ionizes less than 10% in water.
c) HBr is a strong acid, ionizes nearly 100% in water.
d) $H_2C_4H_4O_6$ is a weak acid, ionizes less than 10% in water.

14.43 The term *weak acid* pertains to the amount of dissociation, while the term *dilute* refers to the concentration of the acid in solution.

14.45 Diagram IV is the strongest acid.

SALTS (SEC. 14.6)

14.47 a) HCl and KCl – (1) a salt and an acid b) NaOH and NaCN – (2) a salt and a base
 c) KCl and KBr – (3) two salts d) NH_4Cl and HNO_3 – (1) a salt and an acid

14.49 a) $CaSO_4$, calcium sulfate b) Li_2CO_3, lithium carbonate
 c) NaBr, sodium bromide d) Al_2S_3, aluminum sulfide

14.51 a) $CaSO_4$, insoluble salt b) Li_2CO_3, soluble salt
 c) NaBr, soluble salt d) Al_2S_3, insoluble salt

14.53 a) $NaNO_3$ produces 1 Na^+ and 1 NO_3^-, total 2 ions in water.
 b) $CuCO_3$ produces 1 Cu^{2+} and 1 CO_3^{2-}, total 2 ions in water.
 c) $BaCl_2$ produces 1 Ba^{2+} and 2 Cl^-, total 3 ions in water.
 d) $Al_2(SO_4)_3$ produces 2 Al^{3+} and 3 SO_4^{2-}, total of 5 ions in water.

14.55 In water the following dissociation takes place.
 a) $NaNO_3(s) \rightarrow Na^+(aq) + NO_3^-(aq)$ b) $CuCO_3(s) \rightarrow Cu^{2+}(aq) + CO_3^{2-}(aq)$
 c) $BaCl_2(s) \rightarrow Ba^{2+}(aq) + 2\,Cl^-(aq)$ d) $Al_2(SO_4)_3(s) \rightarrow 2\,Al^{3+}(aq) + 3\,SO_4^{2-}(aq)$

REACTIONS OF ACIDS AND BASES (SECS. 14.7 AND 14.8)

14.57 a) $Fe(s) + 2\,HCl(aq) \rightarrow FeCl_2(aq) + H_2(g)$ b) $Ni(s) + 2\,HCl(aq) \rightarrow NiCl_2(aq) + H_2(g)$
 c) $Ag(s) + HCl(aq) \rightarrow$ No Reaction d) $Pt(s) + HCl(aq) \rightarrow$ No Reaction

14.59 a) Metals that react with cold water: Ca.
 b) Metals that react with hot water (steam): Ca, Zn.
 c) Metals that react with hydrochloric acid (HCl): Zn, Ni, Ca.
 d) Metals that react with H^+ ions to produce H_2 gas: Zn, Ni, Ca.

14.61 a) No, not an acid–base neutralization reaction b) Yes, an acid–base neutralization reaction
 c) Yes, an acid–base neutralization reaction d) No, not an acid–base neutralization reaction

14.63 a) 2:1 b) 1:1 c) 1:1 d) 1:3

14.65 Acid-base neutralization reactions:
 a) $HCN + LiOH \rightarrow LiCN + H_2O$ b) $HCl + NaOH \rightarrow NaCl + H_2O$
 c) $HNO_3 + KOH \rightarrow KNO_3 + H_2O$ d) $HBr + RbOH \rightarrow RbBr + H_2O$

14.67 a) $NaOH + HCl \rightarrow NaCl + H_2O$ b) $KOH + HNO_3 \rightarrow KNO_3 + H_2O$
 c) $2\,LiOH + H_2SO_4 \rightarrow Li_2SO_4 + 2\,H_2O$ d) $Ca(OH)_2 + 2\,HBr \rightarrow CaBr_2 + 2\,H_2O$

14.69 a) $Zn(s) + 2\,HCl(aq) \rightarrow ZnCl_2(aq) + H_2(g)$
 b) $HCl(aq) + NaOH(aq) \rightarrow NaCl(aq) + H_2O(l)$
 c) $2\,HCl(aq) + Na_2CO_3(aq) \rightarrow 2\,NaCl(aq) + CO_2(g) + H_2O(l)$
 d) $HCl(aq) + NaHCO_3(aq) \rightarrow NaCl(aq) + CO_2(g) + H_2O(l)$

REACTIONS OF SALTS (SEC. 14.9)

14.71 a) No, Cu is less reactive than Ni. b) Yes, Zn is more reactive than Ni.
 c) No, Au is less reactive than Ni. d) Yes, Fe is more reactive than Ni.

14.73   a) $Fe(s) + CuSO_4(aq) \rightarrow FeSO_4(aq) + Cu(s)$
   b) $Sn(s) + 2\,AgNO_3(aq) \rightarrow Sn(NO_3)_2(aq) + 2\,Ag(s)$
   c) $Zn(s) + NiCl_2(aq) \rightarrow ZnCl_2(aq) + Ni(s)$
   d) $Cr(s) + Pb(C_2H_3O_2)_2(aq) \rightarrow Cr(C_2H_3O_2)_2(aq) + Pb(s)$

14.75   a) $BaSO_4$, an insoluble salt, is formed.
   b) $Ca_3(PO_4)_2$, an insoluble salt, is formed.
   c) $AgCl$, an insoluble salt is formed, weak acid is formed, $HC_2H_3O_2$.
   d) CO_2, a gas, and H_2O liquid are formed.

14.77   a) $2\,Al(NO_3)_3 + 3\,(NH_4)_2S \rightarrow Al_2S_3 + 6\,NH_4NO_3$ b) $2\,HCl + Ba(OH)_2 \rightarrow 2\,H_2O + BaCl_2$
   c) no reaction   d) no reaction

HYDRONIUM ION AND HYDROXIDE ION CONCENTRATIONS (SEC. 14.10)

14.79 $[H_3O^+] \times [OH^-] = 1.00 \times 10^{-14}$

   a) $5.7 \times 10^{-4} \times [OH^-] = 1.00 \times 10^{-14}$

   $$[OH^-] = \frac{1.00 \times 10^{-14}}{5.7 \times 10^{-4}} = 1.754385965 \times 10^{-11}(calc) = 1.8 \times 10^{-11}\ (corr)$$

   b) $[H_3O^+] \times [6.3 \times 10^{-7}] = 1.00 \times 10^{-14}$

   $$[H_3O^+] = \frac{1.00 \times 10^{-14}}{6.3 \times 10^{-7}} = 1.587301587 \times 10^{-8}\ (calc) = 1.6 \times 10^{-8}\ (corr)$$

   c) $7.70 \times 10^{-10} \times [OH^-] = 1.00 \times 10^{-14}$

   $$[OH^-] = \frac{1.00 \times 10^{-14}}{7.70 \times 10^{-10}} = 1.298701299 \times 10^{-5}(calc) = 1.30 \times 10^{-5}\ (corr)$$

   d) $[H_3O^+] \times [7.03 \times 10^{-12}] = 1.00 \times 10^{-14}$

   $$[H_3O^+] = \frac{1.00 \times 10^{-14}}{7.03 \times 10^{-12}} = 1.422475107 \times 10^{-3}\ (calc) = 1.42 \times 10^{-3}\ (corr)$$

14.81   a) $[H_3O^+] = 3.2 \times 10^{-3}$ acidic
   b) $[H_3O^+] = 5.5 \times 10^{-7}$ acidic (neutral is 1.0×10^{-7})

   c) $[OH^-] = 2.0 \times 10^{-3}, [H_3O^+] = \dfrac{1.00 \times 10^{-14}}{2.0 \times 10^{-3}}$

   $\qquad = 5 \times 10^{-12}\ (calc) = 5.0 \times 10^{-12}\ (corr)$ basic

   d) $[OH^-] = 3.7 \times 10^{-9}, [H_3O^+] = \dfrac{1.00 \times 10^{-14}}{3.7 \times 10^{-9}}$

   $\qquad = 2.702702703 \times 10^{-6}\ (calc) = 2.7 \times 10^{-6}\ (corr)$ acidic

14.83

	$[H_3O^+]$	$[OH^-]$	Acidic or Basic
	2.2×10^{-2}	4.5×10^{-13}	acidic
a)	3.0×10^{-12}	3.3×10^{-3}	basic
b)	6.8×10^{-8}	1.5×10^{-7}	basic
c)	1.4×10^{-7}	7.2×10^{-8}	acidic
d)	4.5×10^{-6}	2.2×10^{-9}	acidic

14.85 at 24°C, $[H_3O^+] \times [OH^-] = 1.00 \times 10^{-14}$
    a) $[H_3O^+] = 3.51 \times 10^{-4}$, $[OH^-] = 2.85 \times 10^{-10}$
      $(3.51 \times 10^{-4}) \times (2.85 \times 10^{-10}) = 1.00035 \times 10^{-13}$ (calc)
                                       $= 1.00 \times 10^{-13}$ (corr) impossible
    b) $[H_3O^+] = 4.72 \times 10^{-7}$, $[OH^-] = 2.12 \times 10^{-8}$
      $(4.72 \times 10^{-7}) \times (2.12 \times 10^{-8}) = 1.00064 \times 10^{-14}$ (calc)
                                         $= 1.00 \times 10^{-14}$ (corr) possible

14.87 At 37°C, the average human body temperature, $K_w = 2.6 \times 10^{-14}$.
    $[H_3O^+] \times [OH^-] = 2.6 \times 10^{-14}$ Since $[H_3O^+] = [OH^-]$,
    $[H_3O^+] = \sqrt{2.6 \times 10^{-14}} = 1.61245155 \times 10^{-7}$ (calc) $= 1.6 \times 10^{-7}$ (corr)
Since $[H_3O^+] = [OH^-]$, the molar hydronium ion concentration, $[H_3O^+]$, and the molar hydroxide ion concentration $[OH^-] = 1.6 \times 10^{-7}$.

THE pH SCALE (SEC. 14.11)

14.89 $pH = -\log[H_3O^+]$
    a) $pH = -\log(1 \times 10^{-4}) = 4$ (calc) $= 4.0$ (corr)
    b) $pH = -\log(1.0 \times 10^{-11}) = 11$ (calc) $= 11.00$ (corr)
    c) $pH = -\log(0.00001) = -\log(1 \times 10^{-5}) = 5$ (calc) $= 5.0$ (corr)
    d) $pH = -\log(0.0000000010) = -\log(1.0 \times 10^{-9}) = 9$ (calc) $= 9.00$ (corr)

14.91 $pH = -\log[H_3O^+]$
    a) $pH = -\log(6.0 \times 10^{-3}) = 2.22184875$ (calc) $= 2.22$ (corr)
    b) $pH = -\log(6.0 \times 10^{-4}) = 3.22184875$ (calc) $= 3.22$ (corr)
    c) $pH = -\log(6.0 \times 10^{-5}) = 4.22184875$ (calc) $= 4.22$ (corr)
    d) $pH = -\log(6.0 \times 10^{-6}) = 5.22184875$ (calc) $= 5.22$ (corr)

14.93 $pH = -\log[H_3O^+]$
    a) $pH = -\log(4 \times 10^{-3}) = 2.397940009$ (calc) $= 2.4$ (corr)
    b) $pH = -\log(4.0 \times 10^{-3}) = 2.397940009$ (calc) $= 2.40$ (corr)
    c) $pH = -\log(4.00 \times 10^{-3}) = 2.397940009$ (calc) $= 2.398$ (corr)
    d) $pH = -\log(4.000 \times 10^{-3}) = 2.397940009$ (calc) $= 2.3979$ (corr)

14.95 a) both acidic   b) acidic, basic   c) both basic   d) acidic, neutral

14.97 a) $pH = -\log[H_3O^+] = 4.0$, therefore antilog $-4.0 = [H_3O^+] = 1 \times 10^{-4}$ M (calc and corr)
    b) $pH = -\log[H_3O^+] = 4.2$, therefore antilog $-4.2 = [H_3O^+] = 6.309573445 \times 10^{-5}$ (calc)
                                           $= 6 \times 10^{-5}$ M (corr)
    c) $pH = -\log[H_3O^+] = 4.5$, therefore antilog $-4.5 = [H_3O^+] = 3.16227766 \times 10^{-5}$ (calc)
                                           $= 3 \times 10^{-5}$ M (corr)
    d) $pH = -\log[H_3O^+] = 4.8$, therefore antilog $-4.8 = [H_3O^+] = 1.584893192 \times 10^{-5}$ (calc)
                                           $= 2 \times 10^{-5}$ M (corr)

14.99 a) $pH = -\log[H_3O^+] = 2.43$, therefore antilog $-2.43 = [H_3O^+] = 3.715352291 \times 10^{-3}$ (calc)
                                           $= 3.7 \times 10^{-3}$ M (corr)
    b) $pH = -\log[H_3O^+] = 5.72$, therefore antilog $-5.72 = [H_3O^+] = 1.905460718 \times 10^{-6}$ (calc)
                                           $= 1.9 \times 10^{-6}$ M (corr)
    c) $pH = -\log[H_3O^+] = 7.73$, therefore antilog $-7.73 = [H_3O^+] = 1.862087137 \times 10^{-8}$ (calc)
                                           $= 1.9 \times 10^{-8}$ M (corr)
    d) $pH = -\log[H_3O^+] = 8.750$, therefore antilog $-8.750 = [H_3O^+] = 1.77827941 \times 10^{-9}$ (calc)
                                           $= 1.78 \times 10^{-9}$ M (corr)

14.101

	$[H_3O^+]$	$[OH^-]$	pH	Acidic or Basic
	6.2×10^{-8}	1.6×10^{-7}	7.21	basic
a)	7.2×10^{-10}	1.4×10^{-5}	9.14	basic
b)	5.0×10^{-6}	2.0×10^{-9}	5.30	acidic
c)	1.4×10^{-5}	7.2×10^{-10}	4.86	acidic
d)	5.9×10^{-9}	1.7×10^{-6}	8.23	basic

14.103 $[H_3O^+]_A = \dfrac{1.0 \times 10^{-14}}{4.3 \times 10^{-4}} = 2.325581395 \times 10^{-11}$ (calc) $= 2.3 \times 10^{-11}$ M (corr)

a) Solution A is more basic. The smaller the $[H_3O^+]$, the more basic the solution.

b) Solution B has the lower pH. The larger the $[H_3O^+]$, the lower the pH.

14.105 The original solution has a $[H_3O^+] = 3.1622777 \times 10^{-5}$ (calc) $= 3.16 \times 10^{-5}$ M(corr)

a) New $[H_3O^+] = 2 \times 3.16 \times 10^{-5} = 6.32 \times 10^{-5}$ M (calc and corr)
   pH $= 4.1992829$ (calc) $= 4.199$ (corr)

b) New $[H_3O^+] = 4 \times 3.16 \times 10^{-5} = 1.264 \times 10^{-4}$ (calc) $= 1.26 \times 10^{-4}$ M (corr)
   pH $= 3.8996295$ (calc) $= 3.900$ (corr)

c) New $[H_3O^+] = 10 \times 3.16 \times 10^{-5} = 3.16 \times 10^{-4}$ M (calc and corr)
   pH $= 3.5003129$ (calc) $= 3.500$ (corr)

d) New $[H_3O^+] = 1000 \times 3.16 \times 10^{-5} = 3.16 \times 10^{-2}$ M (calc and corr)
   pH $= 1.5003129$ (calc) $= 1.500$ (corr)

14.107 a) HNO_3 is a strong acid and ionizes to give 1 H_3O^+ per HNO_3.
   $[H_3O^+] = 6.3 \times 10^{-3}$ M;   pH $= 2.2006595$ (calc) $= 2.20$ (corr)

b) HCl is a strong acid and ionizes to give 1 H_3O^+ per HCl.
   $[H_3O^+] = 0.20$ M;   pH $= 0.6989700$ (calc) $= 0.70$ (corr)

c) H_2SO_4 is a strong acid and ionizes to give 2 H_3O^+ per H_2SO_4.
   $[H_3O^+] = 2 \times 0.000021 = 0.000042$ M;   pH $= 4.3767507$ (calc) $= 4.38$ (corr)

d) NaOH is a strong base containing 1 OH^- per NaOH.
   $[OH^-] = 2.3 \times 10^{-4}$ M;

   $[H_3O^+] = \dfrac{1.00 \times 10^{-14}}{2.3 \times 10^{-4}} = 4.3478261 \times 10^{-11}$ (calc) $= 4.3 \times 10^{-11}$ M (corr)

   pH $= 10.3665315$ (calc) $= 10.37$ (corr)

HYDROLYSIS OF SALTS (SEC. 14.12)

14.109 a) PO_4^{3-} b) CN^- c) NH_4^+ d) none

14.111 a) neutral   b) basic   c) acidic   d) neutral

14.113 a) $NH_4^+ + H_2O \rightarrow H_3O^+ + NH_3$
   b) $C_2H_3O_2^- + H_2O \rightarrow OH^- + HC_2H_3O_2$
   c) $F^- + H_2O \rightarrow OH^- + HF$
   d) $CN^- + H_2O \rightarrow OH^- + HCN$

BUFFERS (SEC. 14.13)

14.115 a) No, have mixture of strong acid and conjugate base
b) Yes, have mixture of weak acid and conjugate base
c) No, do not have an acid in mixture
d) Yes, have mixture of weak acid and conjugate base

14.117 a) HCN and CN^- b) H_3PO_4 and $H_2PO_4^-$ c) H_2CO_3 and HCO_3^- d) HCO_3^- and CO_3^{2-}

14.119 Each diagram contains the weak acid HA and its conjugate base A^-, and therefore, each diagram is considered a buffer solution.

14.121 a) $HF + OH^- \rightarrow F^- + H_2O$ b) $PO_4^{3-} + H_3O^+ \rightarrow HPO_4^{2-} + H_2O$
c) $CO_3^{2-} + H_3O^+ \rightarrow HCO_3^- + H_2O$ d) $H_3PO_4 + OH^- \rightarrow H_2PO_4^- + H_2O$

14.123 Buffer = pH 8.2
a) a small amount of strong acid is added: pH = 8.1
b) a small amount of strong base is added: pH = 8.3

ACID–BASE TITRATIONS (SEC. 14.14)

14.125 a) $HCl + NaOH \rightarrow NaCl + H_2O$

$$20.00\,\text{mL} \times \frac{10^{-3}\,\text{L}}{1\,\text{mL}} \times \frac{0.100\,\text{mole NaOH}}{1\,\text{L}} \times \frac{1\,\text{mole HCl}}{1\,\text{mole NaOH}} \times \frac{1\,\text{L HCl}}{0.100\,\text{mole HCl}}$$
$$\times \frac{1\,\text{mL}}{10^{-3}\,\text{L}} = 20\,(\text{calc}) = 20.0\,\text{mL HCl solution (corr)}$$

b) $HCl + NaOH \rightarrow NaCl + H_2O$

$$20.00\,\text{mL} \times \frac{10^{-3}\,\text{L}}{1\,\text{mL}} \times \frac{0.200\,\text{mole NaOH}}{1\,\text{L}} \times \frac{1\,\text{mole HCl}}{1\,\text{mole NaOH}} \times \frac{1\,\text{L HCl}}{0.100\,\text{mole HCl}}$$
$$\times \frac{1\,\text{mL}}{10^{-3}\,\text{L}} = 40\,(\text{calc}) = 40.0\,\text{mL HCl solution (corr)}$$

c) $2\,HCl + Ba(OH)_2 \rightarrow BaCl_2 + 2\,H_2O$

$$20.00\,\text{mL} \times \frac{10^{-3}\,\text{L}}{1\,\text{mL}} \times \frac{0.100\,\text{mole Ba(OH)}_2}{1\,\text{L}} \times \frac{2\,\text{moles HCl}}{1\,\text{mole Ba(OH)}_2} \times \frac{1\,\text{L HCl}}{0.100\,\text{mole HCl}}$$
$$\times \frac{1\,\text{mL}}{10^{-3}\,\text{L}} = 40\,(\text{calc}) = 40.0\,\text{mL HCl solution (corr)}$$

d) $2\,HCl + Ba(OH)_2 \rightarrow BaCl_2 + 2\,H_2O$

$$20.00\,\text{mL} \times \frac{10^{-3}\,\text{L}}{1\,\text{mL}} \times \frac{0.200\,\text{mole Ba(OH)}_2}{1\,\text{L}} \times \frac{2\,\text{moles HCl}}{1\,\text{mole Ba(OH)}_2} \times \frac{1\,\text{L HCl}}{0.100\,\text{mole HCl}}$$
$$\times \frac{1\,\text{mL}}{10^{-3}\,\text{L}} = 80\,(\text{calc}) = 80.0\,\text{mL HCl solution (corr)}$$

14.127 a) $HCl + KOH \rightarrow KCl + H_2O$

$$20.00\,\text{mL} \times \frac{10^{-3}\,\text{L}}{1\,\text{mL}} \times \frac{0.200\,\text{moles HCl}}{1\,\text{L}} \times \frac{1\,\text{mole KOH}}{1\,\text{mole HCl}} \times \frac{1\,\text{L KOH}}{0.200\,\text{mole KOH}}$$
$$\times \frac{1\,\text{mL}}{10^{-3}\,\text{L}} = 20\,(\text{calc}) = 20.0\,\text{mL KOH solution (corr)}$$

b) $H_2SO_4 + 2\,KOH \rightarrow K_2SO_4 + 2\,H_2O$

$$20.00 \text{ mL} \times \frac{10^{-3}\text{ L}}{1\text{ mL}} \times \frac{0.200 \text{ moles } H_2SO_4}{1\text{ L}} \times \frac{2 \text{ moles KOH}}{1 \text{ mole } H_2SO_4} \times \frac{1 \text{ L KOH}}{0.200 \text{ mole KOH}}$$
$$\times \frac{1 \text{ mL}}{10^{-3}\text{ L}} = 40 \text{ (calc)} = 40.0 \text{ mL KOH solution (corr)}$$

c) $HNO_3 + KOH \rightarrow KNO_3 + H_2O$

$$50.00 \text{ mL} \times \frac{10^{-3}\text{ L}}{1\text{ mL}} \times \frac{0.400 \text{ mole } HNO_3}{1\text{ L}} \times \frac{1 \text{ mole KOH}}{1 \text{ mole } HNO_3} \times \frac{1 \text{ L KOH}}{0.200 \text{ mole KOH}}$$
$$\times \frac{1 \text{ mL}}{10^{-3}\text{ L}} = 100 \text{ mL } HNO_3 \text{ solution (calc)} = 1.00 \times 10^2 \text{ mL } HNO_3 \text{ (corr)}$$

d) $H_2CO_3 + 2\,KOH \rightarrow K_2CO_3 + 2\,H_2O$

$$50.00 \text{ mL} \times \frac{10^{-3}\text{ L}}{1\text{ mL}} \times \frac{0.300 \text{ moles } H_2CO_3}{1\text{ L}} \times \frac{2 \text{ moles KOH}}{1 \text{ mole } H_2CO_3} \times \frac{1 \text{ L KOH}}{0.200 \text{ mole KOH}}$$
$$\times \frac{1 \text{ mL}}{10^{-3}\text{ L}} = 150 \text{ (calc)} = 1.50 \times 10^2 \text{ mL KOH solution (corr)}$$

14.129   a) $H_2SO_4 + 2\,NaOH \rightarrow Na_2SO_4 + 2\,H_2O$

$$34.5 \text{ mL} \times \frac{10^{-3}\text{ L}}{1\text{ mL}} \times \frac{0.102 \text{ mole NaOH}}{1\text{ L}} \times \frac{1 \text{ mole } H_2SO_4}{2 \text{ moles NaOH}} = \frac{1.7595 \times 10^{-3} \text{ mole } H_2SO_4}{0.02500 \text{ L solution}}$$
$$= 7.038 \times 10^{-2} \text{ (calc)} = 0.0704 \text{ M } H_2SO_4 \text{ (corr)}$$

b) $HClO + NaOH \rightarrow NaClO + H_2O$

$$34.5 \text{ mL} \times \frac{10^{-3}\text{ L}}{1\text{ mL}} \times \frac{0.102 \text{ mole NaOH}}{1\text{ L}} \times \frac{1 \text{ mole HClO}}{1 \text{ mole NaOH}} = \frac{3.519 \times 10^{-3} \text{ mole HClO}}{0.02000 \text{ L solution}}$$
$$= 0.17595 \text{ (calc)} = 0.176 \text{ M HClO (corr)}$$

c) $H_3PO_4 + 3\,NaOH \rightarrow Na_3PO_4 + 3\,H_2O$

$$34.5 \text{ mL} \times \frac{10^{-3}\text{ L}}{1\text{ mL}} \times \frac{0.102 \text{ mole NaOH}}{1\text{ L}} \times \frac{1 \text{ mole } H_3PO_4}{3 \text{ moles NaOH}} = \frac{1.173 \times 10^{-3} \text{ mole } H_3PO_4}{0.02000 \text{ L solution}}$$
$$= 5.865 \times 10^{-2} \text{ (calc)} = 0.0587 \text{ M } H_3PO_4 \text{ (corr)}$$

d) $HNO_3 + NaOH \rightarrow NaNO_3 + H_2O$

$$34.5 \text{ mL} \times \frac{10^{-3}\text{ L}}{1\text{ mL}} \times \frac{0.102 \text{ mole NaOH}}{1\text{ L}} \times \frac{1 \text{ mole } HNO_3}{1 \text{ mole NaOH}} = \frac{3.519 \times 10^{-3} \text{ mole } HNO_3}{0.01000 \text{ L solution}}$$
$$= 3.519 \times 10^{-1} \text{ (calc)} = 0.352 \text{ M } HNO_3 \text{ (corr)}$$

Multi-Concept Problems

14.131   **Arrhenius acid**, hydrogen-containing compound that produces H^+ ions in water
        Arrhenius base, hydroxide-ion-containing compound that produces OH^- ions in water
        Bronsted-Lowry acid, substance that can donate H^+
        Bronsted-Lowry base, substance that can accept an H^+
        a) H_3PO_4, Arrhenius acid, Bronsted-Lowry acid
        b) NaOH, Arrhenius base, Bronsted-Lowry base
        c) HNO_3, Arrhenius acid, Bronsted-Lowry acid
        d) Br^-, Bronsted-Lowry base

14.133   a) H_3PO_4 weak acid, triprotic
        b) H_3PO_3 weak acid, triprotic
        c) HBr   strong acid, monoprotic
        d) $HC_2H_3O_2$ weak acid, monoprotic

14.135 a) conjugate of strong acid   b) conjugate of weak acid
c) conjugate of weak acid   d) conjugate of strong acid

14.137 $[H_3O^+]_1 = $ antilog $(-2.2) = 6.3095734 \times 10^{-3}$ (calc) $= 6 \times 10^{-3}$ (corr)
$[H_3O^+]_2 = $ antilog $(-4.5) = 3.1622776 \times 10^{-5}$ (calc) $= 3 \times 10^{-5}$ (corr)
$\dfrac{[H_3O^+]_1}{[H_3O^+]_2} = \dfrac{6 \times 10^{-3}}{3 \times 10^{-5}} = 200$ (calc and corr)

14.139 If we let the hydroxide concentration, $[OH^-] = x$, then $[H_3O^+] = 3x$
From the ion product constant for water, we know that $[H_3O^+] \times [OH^-] = 1.00 \times 10^{-14}$
so $x \times 3x = 1.00 \times 10^{-14}$
giving $x = 5.773503 \times 10^{-8}$
Therefore $[H_3O^+] = 3x = 1.732051 \times 10^{-7}$ and pH $= -\log[H_3O^+] = 6.76144$ (calc) $= 6.76$ (corr).

14.141 a) The lower the pH value, the more acidic the solution. So in decreasing acidity: A, B, D, C.
b) In increasing $[H_3O^+]$: C, D, B, A.
c) The higher the pH value, the more basic the solution. So in decreasing $[OH^-]$: C, D, B, A.
d) In increasing basicity: A, B, D, C.

14.143 a) 0.1 M solution strong acid, pH $= 1$
0.1 M solution weak acid, pH of second solution is higher than 1 (weak acids produce less H^+ in solution than strong acids of same concentration)
b) 0.1 M solution KOH (strong base, pH is high)
0.1 M HNO_3 (acid, pH is low), pH of first solution is higher
c) $[H_3O^+] = 3.0 \times 10^{-3}$, pH $= 2.5222878745$ (calc) $= 2.52$ (corr)
$[H_3O^+] = 3.0 \times 10^{-4}$, pH $= 3.5222878745$ (calc) $= 3.52$ (corr)
pH of second solution is higher, $[H_3O^+]$ is lower
d) $[OH^-] = 6.0 \times 10^{-5}$, $[H_3O^+] = \dfrac{1.00 \times 10^{-14}}{6.0 \times 10^{-5}}$
$= 1.666666667 \times 10^{-10}$ (calc) $= 1.7 \times 10^{-10}$ (corr)
pH $= -\log(1.7 \times 10^{-10}) = 9.769551079$ (calc) $= 9.77$ (corr)
$[OH^-] = 6.0 \times 10^{-6}$, $[H_3O^+] = \dfrac{1.00 \times 10^{-14}}{6.0 \times 10^{-6}}$
$= 1.666666667 \times 10^{-9}$ (calc) $= 1.7 \times 10^{-9}$ (corr)
pH $= -\log(1.7 \times 10^{-9}) = 8.769551079$ (calc) $= 8.77$ (corr)
pH of first solution is higher

14.145 The more concentrated H_3O^+ has a lower pH.
a) no, higher pH   b) no, higher pH   c) yes   d) yes

14.147 NaCl, a soluble salt, does not hydrolyze (no pH effect). HNO_3, a strong acid, produces 0.1 mole H_3O^+ ion. HCl, a strong acid, produces 0.1 mole H_3O^+ ion. NaOH, a strong base, produces 0.1 mole OH^- ion. The 0.1 mole of OH^- ion will neutralize 0.1 mole H_3O^+ ion, with the other 0.1 mole H_3O^+ ion remaining in solution.
$[H_3O^+] = \dfrac{0.1 \text{ mole}}{3.00 \text{ L solution}} = 3.3333333 \times 10^{-2}$ (calc) $= 3 \times 10^{-2}$ M (corr)
pH $= -\log(3 \times 10^{-2}) = 1.5228787$ (calc) $= 1.5$ (corr)

14.149 NH_4Br (salt of a weak base) hydrolyzes to produce a slightly acidic solution. $Ba(OH)_2$ (strong base) produces a strongly basic solution. $HClO_4$ (strong acid) produces a strongly acidic solution. K_2SO_4 (salt of a strong acid and strong base) does not hydrolyze and produces a neutral solution. LiCN (salt of a weak acid) hydrolyzes to produce a slightly basic solution.

        The order of decreasing pH will put the most basic solution first (highest pH) and the most acidic solution last (lowest pH): $Ba(OH)_2$, LiCN, K_2SO_4, NH_4Br, $HClO_4$.

14.151   a)   HCN and NaCN, HCN and KCN   b)   HF and NaF

14.153   Buffer 1: $H_2PO_4^-$, H_3PO_4 Buffer 2: $H_2PO_4^-$, HPO_4^{2-}

14.155   Balanced equation: $3\,Ca(OH)_2 + 2\,H_3PO_4 \rightarrow Ca_3(PO_4)_2 + 6\,H_2O$ Molar mass $H_3PO_4 = 98.00$ g/mol

$$0.40 \text{ g } H_3PO_4 \times \frac{1 \text{ mole } H_3PO_4}{98.00 \text{ g } H_3PO_4} \times \frac{3 \text{ moles } Ca(OH)_2}{2 \text{ moles } H_3PO_4} = 0.00612244898 \text{ (calc)}$$

$$= 0.0061 \text{ mole } Ca(OH)_2 \text{ (corr)}$$

14.157   Balanced equation: $HNO_3 + NaOH \rightarrow NaNO_3 + H_2O$
     Moles $HNO_3 = 0.125$ L $\times$ 5.00 M $= 0.625$ mole
     Moles NaOH $= 0.125$ L $\times$ 6.00 M $= 0.750$ mole
     Therefore after reaction, we have 0.125 mole of NaOH left in the total volume (0.250 L).

$$\text{Concentration of NaOH left is } \frac{0.125 \text{ mole NaOH}}{0.250 \text{ L solution}} = 0.500 \text{ M}$$

     a)   As $[H_3O^+] \times [OH^-] = 1.00 \times 10^{-14}$, $[H_3O^+] = 1.00 \times 10^{-14}/0.50 = 2.00 \times 10^{-14}$ M
     b) $[NaOH] = 0.500$ M
     c)   pH $= -\log[H_3O^+] = 13.69897$ (calc) $= 13.7$ (corr)

Answers to Multiple-Choice Practice Test

MC 14.1 a	**MC 14.2** b	**MC 14.3** b	**MC 14.4** c	**MC 14.5** b	**MC 14.6** b
MC 14.7 c	**MC 14.8** c	**MC 14.9** b	**MC 14.10** b	**MC 14.11** a	**MC 14.12** a
MC 14.13 a	**MC 14.14** c	**MC 14.15** e	**MC 14.16** c	**MC 14.17** e	**MC 14.18** e
MC 14.19 c	**MC 14.20** c				

CHAPTER FIFTEEN
Chemical Equations: Net Ionic and Oxidation–Reduction

ELECTROLYTES (SEC. 15.2)

15.1   a) H_2CO_3 is a weak electrolyte.   b) KOH is a strong electrolyte.
       c) H_2SO_4 is a strong electrolyte.   d) HCN is a weak electrolyte.

15.3   a) Acetic acid, a weak acid, is present in solution in both ionic and molecular forms.
       b) Sucrose, a nonelectrolyte, is a molecular compound present in solution only in a molecular form.
       c) Sodium sulfate, a soluble salt, is present in solution in ionic form.
       d) Hydrofluoric acid, a weak electrolyte, is present in solution in both ionic and molecular forms.

15.5   a) acetic acid, a weak acid   (2) weakly conducting solution
       b) sucrose, a nonelectrolyte   (3) nonconducting solution
       c) sodium sulfate, a soluble salt   (1) strongly conducting solution
       d) hydrofluoric acid, a weak acid   (2) weakly conducting solution

15.7   When dissolved in water:
       a) NaCl produces two ions per formula unit: 1 Na^+ and 1 Cl^-.
       b) $Mg(NO_3)_2$ produces three ions per formula unit: 1 Mg^{2+} and 2 NO_3^-.
       c) NH_4CN produces two ions per formula unit: 1 NH_4^+ and 1 CN^-.
       d) $HClO_4$ produces two ions per formula unit: 1 H^+ and 1 ClO_4^-.

15.9   Diagram III is the strongest electrolyte.

IONIC AND NET IONIC EQUATIONS (SEC. 15.3)

15.11   In an ionic equation:
       a) Strong electrolytes are written in ionic form.
       b) Weak acids are not written in ionic form.
       c) Soluble salts are written in ionic form.
       d) Nonelectrolytes are not written in ionic form.

15.13   In ionic equations:
       a) K_2SO_4, a soluble salt, is written in its ionic form 2 $K^+(aq) + SO_4^{2-}(aq)$.
       b) AgCl, an insoluble salt, is not written in its ionic form.
       c) HCN, a weak acid, is not written in its ionic form.
       d) CO_2, a covalent compound, is not written in an ionic form.

15.15   a) Molecular equation   b) Net ionic equation
       c) Ionic equation   d) Net ionic equation

15.17   a) $2\,NaBr + Pb(NO_3)_2 \rightarrow 2\,NaNO_3 + PbBr_2$

$\quad\quad 2\,Na^+ + 2\,Br^- + Pb^{2+} + 2\,NO_3^- \rightarrow 2\,Na^+ + 2\,NO_3^- + PbBr_2$

$\quad\quad Pb^{2+} + 2\,Br^- \rightarrow PbBr_2$

      b) $FeCl_3 + 3\,NaOH \rightarrow Fe(OH)_3 + 3\,NaCl$

$\quad\quad Fe^{3+} + 3\,Cl^- + 3\,Na^+ + 3\,OH^- \rightarrow Fe(OH)_3 + 3\,Na^+ + 3\,Cl^-$

$\quad\quad Fe^{3+} + 3\,OH^- \rightarrow Fe(OH)_3$

      c) $Zn + 2\,HCl \rightarrow ZnCl_2 + H_2$

$\quad\quad Zn + 2\,H^+ + 2\,Cl^- \rightarrow Zn^{2+} + 2\,Cl^- + H_2$

$\quad\quad Zn + 2\,H^+ \rightarrow Zn^{2+} + H_2$

      d) $H_2S + 2\,KOH \rightarrow K_2S + 2\,H_2O$

$\quad\quad H_2S + 2\,K^+ + 2\,OH^- \rightarrow 2\,K^+ + S^{2-} + 2\,H_2O$

$\quad\quad H_2S + 2\,OH^- \rightarrow S^{2-} + 2\,H_2O$

15.19   a) $Pb + 2\,AgNO_3 \rightarrow 2\,Ag + Pb(NO_3)_2$

$\quad\quad Pb + 2\,Ag^+ + 2\,NO_3^- \rightarrow 2\,Ag + Pb^{2+} + 2\,NO_3^-$

$\quad\quad Pb + 2\,Ag^+ \rightarrow 2\,Ag + Pb^{2+}$

      b) $Cl_2 + 2\,NaBr \rightarrow 2\,NaCl + Br_2$

$\quad\quad Cl_2 + 2\,Na^+ + 2\,Br^- \rightarrow 2\,Na^+ + 2\,Cl^- + Br_2$

$\quad\quad Cl_2 + 2\,Br^- \rightarrow 2\,Cl^- + Br_2$

      c) $2\,Al(NO_3)_3 + 3\,Na_2S \rightarrow Al_2S_3 + 6\,NaNO_3$

$\quad\quad 2\,Al^{3+} + 6\,NO_3^- + 6\,Na^+ + 3\,S^{2-} \rightarrow Al_2S_3 + 6\,Na^+ + 6\,NO_3^-$

$\quad\quad 2\,Al^{3+} + 3\,S^{2-} \rightarrow Al_2S_3$

      d) $NaC_2H_3O_2 + NH_4Cl \rightarrow NH_4C_2H_3O_2 + NaCl$

$\quad\quad Na^+ + C_2H_3O_2^- + NH_4^+ + Cl^- \rightarrow NH_4^+ + C_2H_3O_2^- + Na^+ + Cl^-$

      Each ionic compound is a soluble salt; therefore no reaction occurs.

OXIDATION–REDUCTION TERMINOLOGY (SECS. 15.4 AND 15.5)

15.21   a) Oxidation occurs when an atom loses electrons.

      b) Oxidation occurs when the oxidation number of an atom increases.

15.23   a) An oxidizing agent gains electrons from another substance.

      b) An oxidizing agent contains the atom that shows an oxidation number decrease.

      c) An oxidizing agent is itself reduced.

15.25   a) reducing agent – gain of electrons   incorrect pairing

        causes reduction by providing electrons

      b) substance oxidized – loss of electron   pairing correct

      c) oxidizing agent – increase in oxidation number   incorrect pairing

      d) substance reduced – decrease in oxidation number   pairing correct

OXIDATION NUMBERS (SEC. 15.5)

15.27   a) O_2 – oxidation number of O is 0   b) O_3 – oxidation number of O is 0

      c) H_2O – oxidation number of O is -2 d) CaO – oxidation number of O is -2

15.29   a) $HClO$ – oxidation number of Cl is $+1$ b) $HClO_2$ – oxidation number of Cl is $+3$

      c) $HClO_3$ – oxidation number of Cl is $+5$ d) $HClO_4$ – oxidation number of Cl is $+7$

15.31 a) Br^- – oxidation number of Br is -1 b) IO_4^- – oxidation number of I is $+7$
c) $BeCl_4^{2-}$ – oxidation number of Be is $+2$ d) NF_4^+ – oxidation number of N is $+5$

15.33 a) $CuSO_4$ – oxidation number of Cu is $+2$ b) $Fe_2(SO_4)_3$ – oxidation number of Fe is $+3$
c) $NiSO_4$ – oxidation number of Ni is $+2$ d) Ag_2SO_4 – oxidation number of Ag is $+1$

15.35 a) H_3PO_4, H $= +1$, P $= +5$, O $= -2$
b) $BaCr_2O_7$, Ba $= +2$, Cr $= +6$, O $= -2$
c) NH_4ClO_4, N $= -3$, H $= +1$, Cl $= +7$, O $= -2$
d) $H_4P_2O_7$, H $= +1$, P $= +5$, O $= -2$

15.37 a) $CaCl_2$ – Ca has an oxidation number of $+2$
b) H_2S – no elements have an oxidation number of $+2$
c) OF_2 – no elements have an oxidation number of $+2$
d) $MgBr_2$ – Mg has an oxidation number of $+2$

15.39 a) $Rh_2(CO_3)_2$, $2(Rh) + 2(+4) + 6(-2) = 0$, Rh $= +2$
b) $Cr_2(SO_4)_3$, $2(Cr) + 3(+6) + 12(-2) = 0$, Cr $= +3$
c) $Cu(ClO_2)_2$, Cu $+ 2(+3) + 4(-2) = 0$, Cu $= +2$
d) $Co_3(PO_4)_2$, $3(Co) + 2(+5) + 8(-2) = 0$, Co $= +2$

15.41 a) Na_2O, O $= -2$ b) OF_2, O $= +2$ c) Na_2O_2, O $= -1$ d) BaO, O $= -2$

15.43 a) NaH, H $= -1$ b) CH_4, H $= +1$ c) HCl, H $= +1$ d) CaH_2, H $= -1$

CHARACTERISTICS OF OXIDATION–REDUCTION REACTIONS (SEC. 15.5)

15.45 a) $N_2 + 3H_2 \rightarrow 2NH_3$ Each H in H_2 is oxidized (0 to $+1$); each N in N_2 is reduced (0 to -3).
b) $Cl_2 + 2KI \rightarrow 2KCl + I_2$ I in KI is oxidized (-1 to 0); each Cl in Cl_2 is reduced (0 to -1).
c) $Sb_2O_3 + 3Fe \rightarrow 2Sb + 3FeO$ Fe is oxidized (0 to $+2$); Sb in Sb_2O_3 is reduced ($+3$ to 0).
d) $3H_2SO_3 + 2HNO_3 \rightarrow 2NO + H_2O + 3H_2SO_4$ S in H_2SO_3 is oxidized ($+4$ to $+6$); N in HNO_3 is reduced ($+5$ to $+2$).

15.47 a) $N_2 + 3H_2 \rightarrow 2NH_3$ H_2 is reducing agent; N_2 is oxidizing agent.
b) $Cl_2 + 2KI \rightarrow 2KCl + I_2$ KI is reducing agent; Cl_2 is oxidizing agent.
c) $Sb_2O_3 + 3Fe \rightarrow 2Sb + 3FeO$ Fe is reducing agent; Sb_2O_3 is oxidizing agent.
d) $3H_2SO_3 + 2HNO_3 \rightarrow 2NO + H_2O + 3H_2SO_4$ H_2SO_3 is reducing agent; HNO_3 is oxidizing agent.

15.49 $2HNO_3 + SO_2 \rightarrow H_2SO_4 + 2NO_2$
a) Sulfur in SO_2 is oxidized.
b) HNO_3 is oxidizing agent.
c) HNO_3 contains N, which decreases from $+5$ to $+2$.
d) SO_2 contains S, which loses electrons, $+4$ to $+6$.

REDOX AND NONREDOX CHEMICAL REACTIONS (SEC. 15.6)

15.51 a) $2FeBr_3 \rightarrow 2FeBr_2 + Br_2$, Redox reaction, Fe and Br are changing their oxidation states.
b) $K_2O + H_2O \rightarrow 2KOH$, Nonredox reaction; K, H, and O are not changing their oxidation states.
c) $2KClO_3 \rightarrow 2KCl + 3O_2$, Redox reaction, Cl and O are changing their oxidation states.
d) $CH_4 + 2O_2 \rightarrow CO_2 + 2H_2O$, Redox reaction, C and O are changing their oxidation states.

15.53   a) $H_2 + Cl_2 \rightarrow 2\,HCl$, Synthesis reaction, redox
  b) $Mg + 2\,HBr \rightarrow MgBr_2 + H_2$, Single replacement, redox
  c) $MgCO_3 \rightarrow MgO + CO_2$, Decomposition, nonredox
  d) $2\,KOH + H_2SO_4 \rightarrow H_2SO_4 + 2\,H_2O$, Double replacement (neutralization), nonredox

15.55   a) redox   b) redox   c) cannot classify   d) redox

BALANCING REDOX EQUATIONS: OXIDATION NUMBER METHOD (SEC. 15.8)

15.57   a)   0   2(+3)   +3

$$2\,Cr + 6\,HCl \rightarrow 2\,CrCl_3 + 3\,H_2$$

+1 $3[2(-1)]$ 0

  b)   0   3(+4) +4

$$2\,Cr_2O_3 + 3\,C \rightarrow 4\,Cr + 3\,CO_2$$

+3 $2[2(-3)]$ 0

  c)   +4   (+2)   +6

$$SO_2 + NO_2 \rightarrow SO_3 + NO$$

+4   (−2)   +2

  d)   0   4(+2) +2

$$BaSO_4 + 4\,C \rightarrow BaS + 4\,CO$$

+6   (−8)   −2

15.59   a)   +4   (+2)   +6

$$Br_2 + 2\,H_2O + SO_2 \rightarrow 2\,HBr + H_2SO_4$$

0   2(−1)   −1

  b)   −2   3(+2)   0

$$3\,H_2S + 2\,HNO_3 \rightarrow 3\,S + 2\,NO + 4\,H_2O$$

+5   2(−3)   +2

  c)   +2   2(+1)   +3

$$SnSO_4 + 2\,FeSO_4 \rightarrow Sn + Fe_2(SO_4)_3$$

+2   (−2)   0

d)

$$\overset{-1}{} \qquad \overset{2[2(+1)]}{} \qquad \overset{0}{}$$

$$Na_2TeO_3 + 4\,NaI + 6\,HCl \rightarrow 6\,NaCl + Te + 3\,H_2O + 2\,I_2$$

$$\overset{+4}{} \qquad \overset{(-4)}{} \qquad \overset{0}{}$$

15.61 a)

$$\overset{0}{} \qquad \overset{1[2(+5)]}{} \qquad \overset{+5}{}$$

$$I_2 + 5\,Cl_2 + 6\,H_2O \rightarrow 2\,HIO_3 + 10\,Cl^- + 10\,H^+$$

$$\overset{0}{} \qquad \overset{5[2(1)]}{} \qquad \overset{-1}{}$$

b)

$$\overset{-3}{} \quad \overset{5(+8)}{} \quad \overset{+5}{}$$

$$8\,MnO_4^- + 5\,AsH_3 + 24\,H^+ \rightarrow 5\,H_3AsO_4 + 8\,Mn^{2+} + 12\,H_2O$$

$$\overset{+7}{} \qquad \overset{8(-5)}{} \qquad \overset{+2}{}$$

c)

$$\overset{-1}{} \quad \overset{2(+1)}{} \quad \overset{0}{}$$

$$2\,Br^- + SO_4^{2-} + 4\,H^+ \rightarrow Br_2 + SO_2 + 2\,H_2O$$

$$\overset{+6}{} \quad \overset{(-2)}{} \quad \overset{+4}{}$$

d)

$$\overset{0}{} \qquad \overset{(+3)}{} \qquad \overset{+3}{}$$

$$Au + 4\,Cl^- + 3\,NO_3^- + 6\,H^+ \rightarrow AuCl_4^- + 3\,NO_2 + 3\,H_2O$$

$$\overset{+5}{} \quad \overset{3(-1)}{} \quad \overset{+4}{}$$

15.63 a)

$$\overset{-2}{} \overset{(+8)}{} \overset{+6}{}$$

$$8\,OH^- + S^{2-} + 4\,Cl_2 \rightarrow SO_4^{2-} + 8\,Cl^- + 4\,H_2O$$

$$\overset{0}{} \quad \overset{4[2(-1)]}{} \quad \overset{-1}{}$$

b)

$$\overset{+4}{} \qquad \overset{3(+2)}{} \qquad \overset{+6}{}$$

$$5\,H_2O + 3\,SO_3^{2-} + 2\,CrO_4^{2-} \rightarrow 2\,Cr(OH)_4^- + 3\,SO_4^{2-} + 2\,OH^-$$

$$\overset{+6}{} \quad \overset{2(-3)}{} \quad \overset{+3}{}$$

c)

$$\overset{+5}{} \quad \overset{3(+2)}{} \quad \overset{+7}{}$$

$$H_2O + 2\,MnO_4^- + 3\,IO_3^- \rightarrow 2\,MnO_2 + 3\,IO_4^- + 2\,OH^-$$

$$\overset{+7}{} \quad \overset{2(-3)}{} \quad \overset{+4}{}$$

d)

$$\overset{0}{} \quad \overset{2(+7)}{} \quad \overset{+7}{}$$

$$18\,OH^- + I_2 + 7\,Cl_2 \rightarrow 2\,H_3IO_6^{2-} + 14\,Cl^- + 6\,H_2O$$

$$\overset{0}{} \quad \overset{7[2(-1)]}{} \quad \overset{-1}{}$$

BALANCING REDOX EQUATIONS: HALF-REACTION METHOD (SEC. 15.9)

15.65   a) $NO_3^- \rightarrow NO$, N is reduced from +5 to +3, reduction
       b) $Zn \rightarrow Zn^{2+}$, Zn is oxidized from 0 to +2, oxidation
       c) $Ti^{3+} \rightarrow TiO_2$, Ti is oxidized from +3 to +4, oxidation
       d) $Cr_2O_7^{2-} \rightarrow Cr^{3+}$, Cr is reduced from +6 to +3, reduction

15.67   a) $Te + NO_3^- \rightarrow TeO_2 + NO$, oxidation half-reaction: $Te \rightarrow TeO_2$
       b) $H_2O_2 + Fe^{2+} \rightarrow Fe^{3+} + H_2O$, oxidation half-reaction: $Fe^{2+} \rightarrow Fe^{3+}$
       c) $CN^- + ClO_2^- \rightarrow CNO^- + Cl^-$, oxidation half-reaction: $CN^- \rightarrow CNO^-$
       d) $ClO^- + Cl^- \rightarrow Cl_2$; oxidation half-reaction: $Cl^- \rightarrow Cl_2$

15.69   a) $MnO_2 + e^-$ $\rightarrow Mn^{3+}$ (ox. no.: $+4 \rightarrow +3$)
           $MnO_2 + 4\,H^+ + e^-$ $\rightarrow Mn^{3+}$
           $MnO_2 + 4\,H^+ + e^-$ $\rightarrow Mn^{3+} + 2\,H_2O$

       b) $H_3MnO_4 + 5\,e^-$ $\rightarrow Mn$ (ox. no.: $+5 \rightarrow 0$)
           $H_3MnO_4 + 5\,H^+ + 5\,e^-$ $\rightarrow Mn$
           $H_3MnO_4 + 5\,H^+ + 5\,e^-$ $\rightarrow Mn + 4\,H_2O$

       c) $MnO_4^- + 5\,e^-$ $\rightarrow Mn^{2+}$ (ox. no.: $+7 \rightarrow +2$)
           $MnO_4^- + 8\,H^+ + 5\,e^-$ $\rightarrow Mn^{2+}$
           $MnO_4^- + 8\,H^+ + 5\,e^-$ $\rightarrow Mn^{2+} + 4\,H_2O$

       d) $MnO_4^- + 3\,e^-$ $\rightarrow MnO_2$ (ox. no.: $+7 \rightarrow +4$)
           $MnO_4^- + 4\,H^+ + 3\,e^-$ $\rightarrow MnO_2$
           $MnO_4^- + 4\,H^+ + 3\,e^-$ $\rightarrow MnO_2 + 2\,H_2O$

15.71   a) $SeO_4^{2-} + 6\,e^-$ $\rightarrow Se$ (ox. no.: $+6 \rightarrow 0$)
           $SeO_4^{2-} + 6\,e^-$ $\rightarrow Se + 8\,OH^-$
           $SeO_4^{2-} + 4\,H_2O + 6\,e^-$ $\rightarrow Se + 8\,OH^-$

       b) Se^{2-} $\rightarrow SeO_3^{2-} + 6\,e^-$ (ox. no.: $-2 \rightarrow +4$)
           $Se^{2-} + 6\,OH^-$ $\rightarrow SeO_3^{2-} + 6\,e^-$
           $Se^{2-} + 6\,OH^-$ $\rightarrow SeO_3^{2-} + 6\,e^- + 3\,H_2O$

       c) $SeO_4^{2-} + 2\,e^-$ $\rightarrow SeO_3^{2-}$ (ox. no.: $+6 \rightarrow +4$)
           $SeO_4^{2-} + 2\,e^-$ $\rightarrow SeO_3^{2-} + 2\,OH^-$
           $SeO_4^{2-} + H_2O + 2\,e^-$ $\rightarrow SeO_3^{2-} + 2\,OH^-$

       d) Se $\rightarrow SeO_3^{2-} + 4\,e^-$ (ox. no.: $0 \rightarrow +4$)
           $Se + 6\,OH^-$ $\rightarrow SeO_3^{2-} + 4\,e^-$
           $Se + 6\,OH^-$ $\rightarrow SeO_3^{2-} + 4\,e^- + 3\,H_2O$

15.73   a) $Fe^{3+} + 1\,e^- \rightarrow Fe^{2+}$
       b) $Cr \rightarrow Cr^{3+} + 3\,e^-$
       c) $Zn + 2\,OH^- \rightarrow Zn(OH)_2 + 2\,e^-$
       d) $S + 2\,H^+ + 2\,e^- \rightarrow H_2S$

15.75   a) oxidation:
          Zn $\rightarrow Zn^{2+} + 2\,e^-$ (ox. no.: $0 \rightarrow +2$)
       reduction:
          $Cu^{2+} + 2\,e^-$ $\rightarrow Cu$ (ox. no.: $+2 \rightarrow 0$)
       combining:
          $Zn \rightarrow Zn^{2+} + 2\,e^-$ $+$ $Cu^{2+} + 2\,e^- \rightarrow Cu$
          $Zn + Cu^{2+} + 2\,e^-$ $\rightarrow Zn^{2+} + 2\,e^- + Cu$
          $Zn + Cu^{2+}$ $\rightarrow Zn^{2+} + Cu$

b) oxidation:

$$2\,I^- \quad\rightarrow I_2 + 2\,e^- \;(\text{ox. no.: } -1 \rightarrow 0)$$

reduction:

$$Br_2 + 2\,e^- \quad\rightarrow 2\,Br^-\,(\text{ox. no.: } 0 \rightarrow -1)$$

combining:

$$2\,I^- \rightarrow I_2 + 2\,e^- \quad + \quad Br_2 + 2\,e^- \rightarrow 2\,Br^-$$
$$2\,I^- + Br_2 + 2\,e^- \quad\rightarrow I_2 + 2\,e^- + 2\,Br^-$$
$$2\,I^- + Br_2 \quad\rightarrow I_2 + 2\,Br^-$$

c) oxidation:

$$S_2O_3{}^{2-} \quad\rightarrow 2\,HSO_4{}^-$$
$$S_2O_3{}^{2-} \quad\rightarrow 2\,HSO_4{}^- + 8\,e^-\;(\text{ox. no.: } +2 \rightarrow +6\text{ per S atom})$$
$$S_2O_3{}^{2-} \quad\rightarrow 2\,HSO_4{}^- + 8\,e^- + 8\,H^+$$
$$S_2O_3{}^{2-} + 5\,H_2O \quad\rightarrow 2\,HSO_4{}^- + 8\,e^- + 8\,H^+$$

reduction:

$$Cl_2 + 2\,e^- \quad\rightarrow 2\,Cl^-\;(\text{ox. no.: } 0 \rightarrow -1)$$

combining:

$$S_2O_3{}^{2-} + 5\,H_2O \rightarrow 2\,HSO_4{}^- + 8\,e^- + 8\,H^+ \quad + \quad 4(Cl_2 + 2\,e^- \rightarrow 2\,Cl^-)$$
$$S_2O_3{}^{2-} + 5\,H_2O + 4\,Cl_2 + 8\,e^- \quad\rightarrow 2\,H_2SO_4{}^- + 8\,e^- + 8\,H^+ + 8\,Cl^-$$
$$S_2O_3{}^{2-} + 5\,H_2O + 4\,Cl_2 \quad\rightarrow 2\,HSO_4{}^- + 8\,H^+ + 8\,Cl^-$$

d) oxidation:

$$Zn \quad\rightarrow Zn^{2+} + 2\,e^-\;(\text{ox. no.: } 0 \rightarrow +2)$$

reduction:

$$As_2O_3 \quad\rightarrow 2\,AsH_3\;(\text{balance As})$$
$$As_2O_3 + 12\,e^- \quad\rightarrow 2\,AsH_3\;(\text{ox. no.: } +3 \rightarrow -3\text{ per As atom})$$
$$As_2O_3 + 12\,e^- + 12\,H^+ \quad\rightarrow 2\,AsH_3$$
$$As_2O_3 + 12\,e^- + 12\,H^+ \quad\rightarrow 2\,AsH_3 + 3\,H_2O$$

combining:

$$6(Zn \rightarrow Zn^{2+} + 2e) \quad + \quad As_2O_3 + 12\,e^- + 12\,H^+$$
$$\rightarrow 2\,AsH_3 + 3\,H_2O$$
$$6\,Zn + As_2O_3 + 12\,e^- + 12\,H^+ \quad\rightarrow 6\,Zn^{2+} + 12\,e^- + 2\,AsH_3 + 3\,H_2O$$
$$6\,Zn + As_2O_3 + 12\,H^+ \quad\rightarrow 6\,Zn^{2+} + 2\,AsH_3 + 3\,H_2O$$

15.77 a) oxidation:

$$2\,NH_3 \quad\rightarrow N_2H_4\;(\text{balance N})$$
$$2\,NH_3 \quad\rightarrow N_2H_4 + 2\,e^-\;(\text{ox. no.: } -3 \rightarrow -2\text{ per N atom})$$
$$2\,NH_3 + 2\,OH^- \quad\rightarrow N_2H_4 + 2\,e^-$$
$$2\,NH_3 + 2\,OH^- \quad\rightarrow N_2H_4 + 2\,e^- + 2\,H_2O$$

reduction:

$$ClO^- + 2\,e^- \quad\rightarrow Cl^-\;(\text{ox. no.: } +1 \rightarrow -1)$$
$$ClO^- + 2\,e^- \quad\rightarrow Cl^- + 2\,OH^-$$
$$ClO^- + 2\,e^- + H_2O \quad\rightarrow Cl^- + 2\,OH^-$$

combining:

$$2\,NH_3 + 2\,OH^- \rightarrow N_2H_4 + 2\,e^- + 2\,H_2O \;+\; ClO^- + 2\,e^- + H_2O \rightarrow Cl^- + 2\,OH^-$$
$$2\,NH_3 + 2\,OH^- + ClO^- + 2\,e^- + H_2O \quad\rightarrow N_2H_4 + 2\,e^- + 2\,H_2O + Cl^- + 2\,OH^-$$
$$2\,NH_3 + ClO^- \quad\rightarrow N_2H_4 + Cl^- + H_2O$$

b) oxidation:

$$Cr(OH)_2 \quad\rightarrow CrO_4{}^{2-} + 4\,e^-\;(\text{ox. no.: } +2 \rightarrow +6)$$
$$Cr(OH)_2 + 6\,OH^- \quad\rightarrow Br^- \; CrO_4{}^{2-} + 4\,e^-$$
$$Cr(OH)_2 + 6\,OH^- \quad\rightarrow CrO_4{}^{2-} + 4\,H_2O + 4\,e^-$$

reduction:

$$BrO^- + 2\,e^- \qquad\qquad \rightarrow Br^- \text{ (ox. no.: } +1 \rightarrow -1)$$
$$BrO^- + 2\,e^- \qquad\qquad \rightarrow Br^- + 2\,OH^-$$
$$BrO^- + H_2O + 2\,e^- \qquad \rightarrow Br^- + 2\,OH^-$$

combining:

$$Cr(OH)_2 + 6\,OH^- \rightarrow CrO_4^- + 4\,H_2O + 4\,e^- \quad + \quad 2(BrO^- + H_2O + 2\,e^- \rightarrow Br^- + 2\,OH^-)$$
$$Cr(OH)_2 + 6\,OH^- + 2\,BrO^- + 2\,H_2O + 4\,e^-$$
$$\qquad\qquad\qquad\qquad\qquad \rightarrow CrO_4^{2-} + 4\,H_2O + 4\,e^- + 2\,Br^- + 4\,OH^-$$
$$Cr(OH)_2 + 2\,OH^- + 2\,BrO^- \qquad \rightarrow CrO_4^{2-} + 2\,H_2O + 2\,Br^-$$

c) oxidation:

$$CrO_2^- \qquad\qquad\qquad \rightarrow CrO_4^{2-} + 3\,e^- \text{ (ox. no.: } +3 \rightarrow +6)$$
$$CrO_2^- + 4\,OH^- \qquad\qquad \rightarrow CrO_4^{2-} + 3\,e^-$$
$$CrO_2^- + 4\,OH^- \qquad\qquad \rightarrow CrO_4^{2-} + 3\,e^- + 2\,H_2O$$

reduction:

$$H_2O_2 + 2\,e^- \qquad\qquad \rightarrow 2\,OH^- \text{ (ox. no.: } -1 \rightarrow -2 \text{ per O atom)}$$

combining:

$$2(CrO_2^- + 4\,OH^- \rightarrow CrO_4^{2-} + 3\,e^- + 2\,H_2O) \quad + \quad 3(H_2O_2 + 2\,e^- \rightarrow 2\,OH^-)$$
$$2\,CrO_2^- + 8\,OH^- + 3\,H_2O_2 + 6\,e^- \qquad \rightarrow 2\,CrO_4^{2-} + 6\,e^- + 4\,H_2O + 6\,OH^-$$
$$2\,CrO_2^- + 2\,OH^- + 3\,H_2O_2 \qquad \rightarrow 2\,CrO_4^{2-} + 4\,H_2O$$

d) oxidation:

$$Sn(OH)_3^- \qquad\qquad \rightarrow Sn(OH)_6^{2-} + 2\,e^- \text{ (ox. no.: } +2 \rightarrow +4)$$
$$Sn(OH)_3^- + 3\,OH^- \qquad \rightarrow Sn(OH)_6^{2-} + 2\,e^-$$

reduction:

$$Bi(OH)_3 + 3\,e^- \qquad\qquad \rightarrow Bi \text{ (ox. no.: } +3 \rightarrow 0)$$
$$Bi(OH)_3 + 3\,e^- \qquad\qquad \rightarrow Bi + 3\,OH^-$$

combining:

$$3(Sn(OH)_3^- + 3\,OH^- \rightarrow Sn(OH)_6^{2-} + 2\,e^-) \quad + \quad 2(Bi(OH)_3 + 3\,e^- \rightarrow Bi + 3\,OH^-)$$
$$3\,Sn(OH)_3^- + 9\,OH^- + 2\,Bi(OH)_3 + 6\,e^- \rightarrow 3\,Sn(OH)_6^{2-} + 6\,e^- + 2\,Bi + 6\,OH^-$$
$$3\,Sn(OH)_3^- + 3\,OH^- + 2\,Bi(OH)_3 \qquad \rightarrow 3\,Sn(OH)_6^{2-} + 2\,Bi$$

15.79 a) $Pb + 2\,Cl^- \rightarrow PbCl_2 + 2\,e^-$ b) $Sn(OH)_3^- + 3\,OH^- \rightarrow Sn(OH)_6^{2-} + 2\,e^-$
c) $BiO_3 + 6\,H^+ + 3\,e^- \rightarrow Bi^{3+} + 3\,H_2O$ d) $SO_4^{2-} + 9\,H^+ + 8\,e^- \rightarrow HS^- + 4\,H_2O$

15.81 reduction: $3(O_2 + 4\,H^+ + 4\,e^- \rightarrow 2\,H_2O)$ 3×4 electrons transferred
$$3\,O_2 + 12\,H^+ + 12\,e^- \rightarrow 6\,H_2O$$
oxidation: $2\,(PbS + 3\,H_2O \rightarrow PbO + SO_2 + 6\,e^- + 6\,H^+)$
$$2 \times 6 \text{ electrons transferred}$$
$$2\,PbS + 6\,H_2O \rightarrow 2\,PbO + 2\,SO_2 + 12\,e^- + 12\,H^+$$
combining ½ reactions $=$
$$3\,O_2 + 12\,H^+ + 2\,PbS + 6\,H_2O + 12\,e^- \rightarrow 6\,H_2O + 2\,PbO + 2\,SO_2 + 12\,e^- + 12\,H^+$$
overall reaction $= 2\,PbS + 3\,O_2 \rightarrow 2\,PbO + 2\,SO_2$

BALANCING REDOX REACTIONS: DISPROPORTIONATION REACTIONS (SEC. 15.10)

15.83 a)
$$\begin{array}{c} +3 \quad (+2) \quad +5 \\ \overline{} \\ 2\,HNO_2 + HNO_2 \rightarrow 2\,NO + NO_3^- + H_2O + H^+ \\ \underline{} \\ +3 \quad 2(-1) \quad +2 \end{array}$$

$$3\,HNO_2 \rightarrow 2\,NO + NO_3^- + H_2O + H^+$$

b)
$$\overset{\displaystyle -1 \qquad 2(+1) \qquad\quad 0}{2\,Cl^- + 2\,ClO^- + 4\,H^+ \rightarrow Cl_2 + Cl_2 + 2\,H_2O}$$
$$+1 \qquad 2(-1) \qquad\quad 0$$

$$2\,Cl^- + 2\,ClO^- + 4\,H^+ \rightarrow 2\,Cl_2 + 2\,H_2O = Cl^- + ClO^- + 2\,H^+ \rightarrow Cl_2 + H_2O$$

c)
$$\overset{\displaystyle 0 \quad (+4) \;\; +4}{6\,OH^- + 2\,S + S \rightarrow 2\,S^{2-} + SO_3^{2-} + 3\,H_2O}$$
$$0 \;\; 2(-2) \;\; -2$$

$$6\,OH^- + 3\,S \rightarrow 2\,S^{2-} + SO_3^{2-} + 3\,H_2O$$

d)
$$\overset{\displaystyle 0 \quad 2(+5) \;\; +5}{12\,OH^- + Br_2 + 5\,Br_2 \rightarrow 2\,BrO_3^- + 10\,Br^- + 6\,H_2O}$$
$$0 \quad 5[2(-1)] \;\; -1$$

$$12\,OH^- + 6\,Br_2 \rightarrow 2\,BrO_3^- + 10\,Br^- + 6\,H_2O = 6\,OH^- + 3\,Br_2 \rightarrow BrO_3^- + 5\,Br^- + 3\,H_2O$$

15.85 a) oxidation:

HNO_2	$\rightarrow NO_3^- + 2\,e^-$
HNO_2	$\rightarrow NO_3^- + 2\,e^- + 3\,H^+$
$HNO_2 + H_2O$	$\rightarrow NO_3^- + 2\,e^- + 3\,H^+$

reduction:

$HNO_2 + e^-$	$\rightarrow NO$
$HNO_2 + e^- + H^+$	$\rightarrow NO$
$HNO_2 + e^- + H^+$	$\rightarrow NO + H_2O$

combining:

$HNO_2 + H_2O \rightarrow NO_3^- + 3\,H^+ + 2\,e^-$ **+**	$2(HNO_2 + e^- + H^+ \rightarrow NO + H_2O)$
$HNO_2 + H_2O + 2\,HNO_2 + 2\,H^+ + 2\,e^-$	$\rightarrow NO_3^- + 2\,e^- + 3\,H^+ + 2\,NO + 2\,H_2O$
$3\,HNO_2$	$\rightarrow NO_3^- + 2\,NO + H^+ + H_2O$

b) oxidation:

$2\,Cl^-$	$\rightarrow Cl_2$
$2\,Cl^-$	$\rightarrow Cl_2 + 2\,e^-$

reduction:

$2\,ClO^-$	$\rightarrow Cl_2$
$2\,ClO^- + 2\,e^-$	$\rightarrow Cl_2$
$2\,ClO^- + 2\,e^- + 4\,H^+$	$\rightarrow Cl_2$
$2\,ClO^- + 2\,e^- + 4\,H^+$	$\rightarrow Cl_2 + 2\,H_2O$

combining:

$2\,Cl^- \rightarrow Cl_2 + 2\,e^-$ **+** $2\,ClO^- + 2\,e^- + 4\,H^+ \rightarrow Cl_2 + 2\,H_2O$	
$2\,Cl^- + 2\,ClO^- + 2\,e^- + 4\,H^+$	$\rightarrow Cl_2 + 2\,e^- + Cl_2 + 2\,H_2O$
$2\,Cl^- + 2\,ClO^- + 4\,H^+$	$\rightarrow 2\,Cl_2 + 2\,H_2O$
$Cl^- + ClO^- + 2\,H^+$	$\rightarrow Cl_2 + H_2O$

c) oxidation:

$$S \rightarrow SO_3^{2-} + 4\,e^-$$
$$S + 6\,OH^- \rightarrow SO_3^{2-} + 4\,e^-$$
$$S + 6\,OH^- \rightarrow SO_3^{2-} + 4\,e^- + 3\,H_2O$$

reduction:

$$S + 2\,e^- \rightarrow S^{2-}$$

combining:

$$S + 6\,OH^- \rightarrow SO_3^{2-} + 4\,e^- + 3\,H_2O + 2(S + 2\,e^- \rightarrow S^{2-})$$
$$S + 6\,OH^- + 2\,S + 4\,e^- \rightarrow SO_3^{2-} + 4\,e^- + 2\,S^{2-} + 3\,H_2O$$
$$3\,S + 6\,OH^- \rightarrow SO_3^{2-} + 2\,S^{2-} + 3\,H_2O$$

d) oxidation:

$$Br_2 \rightarrow 2\,BrO_3^-$$
$$Br_2 \rightarrow 2\,BrO_3^- + 10\,e^-$$
$$Br_2 + 12\,OH^- \rightarrow 2\,BrO_3^- + 10\,e^-$$
$$Br_2 + 12\,OH^- \rightarrow 2\,BrO_3^- + 10\,e^- + 6\,H_2O$$

reduction:

$$Br_2 + 2\,e^- \rightarrow 2\,Br^-$$

combining:

$$Br_2 + 12\,OH^- \rightarrow 2\,BrO_3^- + 10\,e^- + 6\,H_2O \quad + \quad 5(Br_2 + 2\,e^- \rightarrow 2\,Br^-)$$
$$Br_2 + 12\,OH^- + 5\,Br_2 + 10\,e^- \rightarrow 2\,BrO_3^- + 10\,e^- + 6\,H_2O + 10\,Br^-$$
$$6\,Br_2 + 12\,OH^- \rightarrow 2\,BrO_3^- + 10\,Br^- + 6\,H_2O$$
$$3\,Br_2 + 6\,OH^- \rightarrow BrO_3^- + 5\,Br^- + 3\,H_2O$$

STOICHIOMETRIC CALCULATIONS INVOLVING IONS (SEC. 15.11)

15.87 a) $Na = 22.99$ g/mole

b) $Na^+ = 22.99$ g/mole

c) $SO_3 = 1\,S = 32.06$ g/mole

$\underline{+3\,O = 48.00 \text{ g/mole}}$

80.06 g/mole

d) $SO_3^{2-} = 1\,S = 32.06$ g/mole

$\underline{+3\,O = 48.00 \text{ g/mole}}$

80.06 g/mole

15.89 a) $1.00 \times 10^{24} \text{ CO}_3^{2-} \text{ ions} \times \dfrac{1 \text{ mole CO}_3^{2-} \text{ ions}}{6.022 \times 10^{23} \text{ CO}_3^{2-} \text{ ions}} \times \dfrac{1 \text{ mole Mg}^{2+} \text{ ions}}{1 \text{ mole CO}_3^{2-} \text{ ions}} \times \dfrac{24.31 \text{ g Mg}^{2+} \text{ ions}}{1 \text{ mole Mg}^{2+} \text{ ions}}$

$= 40.36864829 \text{ (calc)} = 40.4 \text{ g Mg}^{2+} \text{ (corr)}$

b) $1.00 \times 10^{24} \text{ SO}_4^{2-} \text{ ions} \times \dfrac{1 \text{ mole SO}_4^{2-} \text{ ions}}{6.022 \times 10^{23} \text{ SO}_4^{2-} \text{ ions}} \times \dfrac{1 \text{ mole Mg}^{2+} \text{ ions}}{1 \text{ mole SO}_4^{2-} \text{ ions}} \times \dfrac{24.31 \text{ g Mg}^{2+} \text{ ions}}{1 \text{ mole Mg}^{2+} \text{ ions}}$

$= 40.36864829 \text{ (calc)} = 40.4 \text{ g Mg}^{2+} \text{ (corr)}$

c) $1.00 \times 10^{24} \text{ Cl}^- \text{ ions} \times \dfrac{1 \text{ mole Cl}^- \text{ ions}}{6.022 \times 10^{23} \text{ Cl}^- \text{ ions}} \times \dfrac{1 \text{ mole Mg}^{2+} \text{ ions}}{2 \text{ moles Cl}^- \text{ ions}} \times \dfrac{24.31 \text{ g Mg}^{2+} \text{ ions}}{1 \text{ mole Mg}^{2+} \text{ ions}}$

$= 20.18432414 \text{ (calc)} = 20.2 \text{ g Mg}^{2+} \text{ (corr)}$

d) $1.00 \times 10^{24} \text{ total ions} \times \dfrac{1 \text{ mole ions}}{6.022 \times 10^{23} \text{ ions}} \times \dfrac{1 \text{ mole Mg}^{2+} \text{ ions}}{3 \text{ moles ions}} \times \dfrac{24.31 \text{ g Mg}^{2+} \text{ ions}}{1 \text{ mole Mg}^{2+} \text{ ions}}$

$= 13.4562161 \text{ (calc)} = 13.5 \text{ g Mg}^{2+} \text{ (corr)}$

15.91 a) $0.634 \text{ mole Ca}^{2+} \text{ ion} \times \dfrac{1 \text{ mole Ca(CN)}_2}{1 \text{ mole Ca}^{2+} \text{ ions}} \times \dfrac{1 \text{ L solution}}{0.500 \text{ mole Ca(CN)}_2} \times \dfrac{1 \text{ mL}}{10^{-3} \text{ L}}$

$$= 1268 \text{ (calc)} = 1270 \text{ mL (corr)}$$

b) $0.634 \text{ mole CN}^- \text{ ion} \times \dfrac{1 \text{ mole Ca(CN)}_2}{2 \text{ moles CN}^- \text{ ions}} \times \dfrac{1 \text{ L solution}}{0.500 \text{ mole Ca(CN)}_2} \times \dfrac{1 \text{ mL}}{10^{-3} \text{ L}}$

$$= 634 \text{ mL (calc and corr)}$$

c) $0.553 \text{ mole Ca}^{2+} \text{ ion} \times \dfrac{1 \text{ mole Ca(CN)}_2}{1 \text{ mole Ca}^{2+} \text{ ions}} \times \dfrac{1 \text{ L solution}}{0.500 \text{ mole Ca(CN)}_2} \times \dfrac{1 \text{ mL}}{10^{-3} \text{ L}}$

$$= 1106 \text{ (calc)} = 1110 \text{ mL (corr)}$$

d) $1.20 \text{ total moles ions} \times \dfrac{1 \text{ mole Ca(CN)}_2}{3 \text{ moles ions}} \times \dfrac{1 \text{ L solution}}{0.500 \text{ mole Ca(CN)}_2} \times \dfrac{1 \text{ mL}}{10^{-3} \text{ L}}$

$$= 800. \text{ mL (calc and corr)}$$

15.93 $20.0 \text{ g Fe} \times \dfrac{1 \text{ mole Fe}}{55.84 \text{ g Fe}} \times \dfrac{1 \text{ mole Cu}^{2+}}{1 \text{ mole Fe}} \times \dfrac{1 \text{ L solution}}{1.35 \text{ moles Cu}^{2+}} \times \dfrac{1 \text{ mL}}{10^{-3} \text{ L}}$

$$= 265.3082882 \text{ (calc)} = 265 \text{ mL (corr)}$$

15.95 a) $0.20 \text{ M NaCl} = 1(0.20 \text{ M}) \text{ Na}^+ = 0.20 \text{ M Na}^+$ (calc and corr)

$\qquad 1(0.20 \text{ M}) \text{ Cl}^- = 0.20 \text{ M Cl}^-$ (calc and corr)

b) $0.20 \text{ M K}_2\text{SO}_4 = 2(0.20 \text{ M}) \text{ K}^+ = 0.40 \text{ M K}^+$ (calc and corr)

$\qquad 1(0.20 \text{ M}) \text{ SO}_4^{2-} = 0.20 \text{ M SO}_4^{2-}$ (calc and corr)

c) $0.20 \text{ M Al(NO}_3)_3 = 1(0.20 \text{ M}) \text{ Al}^{3+} = 0.20 \text{ M Al}^{3+}$ (calc and corr)

$\qquad 3(0.20 \text{ M}) \text{ NO}_3^- = 0.60 \text{ M NO}_3^-$ (calc and corr)

d) $0.20 \text{ M MgCl}_2 = 1(0.20 \text{ M}) \text{ Mg}^{2+} = 0.20 \text{ M Mg}^{2+}$ (calc and corr)

$\qquad 2(0.20 \text{ M}) \text{ Cl}^- = 0.40 \text{ M Cl}^-$ (calc and corr)

15.97 a) $8.45 \text{ g NaBr} \times \dfrac{1 \text{ mole NaBr}}{102.89 \text{ g NaBr}} \times \dfrac{2 \text{ moles ions}}{1 \text{ mole NaBr}} \times \dfrac{6.022 \times 10^{23} \text{ ions}}{1 \text{ mole ions}}$

$$= 9.891320828 \times 10^{22} \text{ (calc)} = 9.89 \times 10^{22} \text{ ions (corr)}$$

b) $3.20 \text{ g K}_2\text{SO}_4 \times \dfrac{1 \text{ mole K}_2\text{SO}_4}{174.27 \text{ g K}_2\text{SO}_4} \times \dfrac{3 \text{ moles ions}}{1 \text{ mole K}_2\text{SO}_4} \times \dfrac{6.022 \times 10^{23} \text{ ions}}{1 \text{ mole ions}}$

$$= 3.31733517 \times 10^{22} \text{ (calc)} = 3.32 \times 10^{22} \text{ ions (corr)}$$

c) $30.0 \text{ g HCl} \times \dfrac{1 \text{ mole HCl}}{36.46 \text{ g HCl}} \times \dfrac{2 \text{ moles ions}}{1 \text{ mole HCl}} \times \dfrac{6.022 \times 10^{23} \text{ ions}}{1 \text{ mole ions}}$

$$= 9.910038398 \times 10^{23} \text{ (calc)} = 9.91 \times 10^{23} \text{ ions (corr)}$$

d) $40.0 \text{ g KOH} \times \dfrac{1 \text{ mole KOH}}{56.11 \text{ g KOH}} \times \dfrac{2 \text{ moles ions}}{1 \text{ mole KOH}} \times \dfrac{6.022 \times 10^{23} \text{ ions}}{1 \text{ mole ions}}$

$$= 8.585991802 \times 10^{23} \text{ (calc)} = 8.59 \times 10^{23} \text{ ions (corr)}$$

15.99 a) $10.0 \text{ g Na}_2\text{SO}_4 \times \dfrac{1 \text{ mole Na}_2\text{SO}_4}{142.05 \text{ g Na}_2\text{SO}_4} \times \dfrac{1 \text{ L solution}}{0.125 \text{ mole Na}_2\text{SO}_4} \times \dfrac{1 \text{ mL}}{10^{-3} \text{ L}}$

$$= 563.1819782 \text{ (calc)} = 563 \text{ mLs (corr)}$$

b) $2.5 \text{ g Na}^+ \times \dfrac{1 \text{ mole Na}^+}{22.99 \text{ g Na}^+} \times \dfrac{1 \text{ mole Na}_2\text{SO}_4}{2 \text{ moles Na}^+} \times \dfrac{1 \text{ L solution}}{0.125 \text{ mole Na}_2\text{SO}_4} \times \dfrac{1 \text{ mL}}{10^{-3} \text{ L}}$

$$= 434.9717268 \text{ (calc)} = 4.3 \times 10^2 \text{ mLs (corr)}$$

c) $0.567 \text{ mole Na}_2\text{SO}_4 \times \dfrac{1 \text{ L solution}}{0.125 \text{ mole Na}_2\text{SO}_4} \times \dfrac{1 \text{ mL}}{10^{-3} \text{ L}} = 4536 \text{ (calc)} = 4540 \text{ mLs (corr)}$

d) $0.112 \text{ mole SO}_4{}^{2-} \times \dfrac{1 \text{ mole Na}_2\text{SO}_4}{1 \text{ mole SO}_4{}^{2-}} \times \dfrac{1 \text{ L solution}}{0.125 \text{ mole Na}_2\text{SO}_4} \times \dfrac{1 \text{ mL}}{10^{-3} \text{ L}} = 896 \text{ mLs (calc and corr)}$

15.101 a) $27 \text{ mL} \times \dfrac{10^{-3} \text{ L}}{1 \text{ mL}} \times \dfrac{0.200 \text{ mole KCl}}{1 \text{ L solution}} \times \dfrac{1 \text{ mole K}^+}{1 \text{ mole KCl}} = 0.0054 \text{ mole K}^+ \text{ (calc and corr)}$

    $175 \text{ mL} \times \dfrac{10^{-3} \text{ L}}{1 \text{ mL}} \times \dfrac{0.100 \text{ mole K}_3\text{PO}_4}{1 \text{ L solution}} \times \dfrac{3 \text{ mole K}^+}{1 \text{ mole K}_3\text{PO}_4} = 0.0525 \text{ mole K}^+ \text{ (calc and corr)}$

    Total moles of $\text{K}^+ = 0.0525 + 0.0054 = 0.0579 \text{ mole K}^+$

    Concentration of $\text{K}^+ = \dfrac{0.0579 \text{ mole K}^+}{(175 + 27) \text{ mL}} \times \dfrac{1 \text{ mL}}{10^{-3} \text{ L}} = 0.286633663 \text{ (calc)} = 0.287 \text{ M K}^+ \text{ (corr)}$

 b) $27 \text{ mL} \times \dfrac{10^{-3} \text{ L}}{1 \text{ mL}} \times \dfrac{0.200 \text{ mole KCl}}{1 \text{ L solution}} \times \dfrac{1 \text{ mole Cl}^-}{1 \text{ mole KCl}} = 0.0054 \text{ mole Cl}^- \text{ (calc and corr)}$

    Concentration of $\text{Cl}^- = \dfrac{0.0054 \text{ mole Cl}^-}{(175 + 27) \text{ mL}} \times \dfrac{1 \text{ mL}}{10^{-3} \text{ L}} = 0.02673267 \text{ (calc)} = 0.027 \text{ M Cl}^- \text{ (corr)}$

 c) $175 \text{ mL} \times \dfrac{10^{-3} \text{ L}}{1 \text{ mL}} \times \dfrac{0.100 \text{ mole K}_3\text{PO}_4}{1 \text{ L solution}} \times \dfrac{1 \text{ mole PO}_4{}^{3-}}{1 \text{ mole K}_3\text{PO}_4} = 0.0175 \text{ mole PO}_4{}^{3-} \text{ (calc and corr)}$

    Concentration of $\text{PO}_4{}^{3-} = \dfrac{0.0175 \text{ mole PO}_4{}^{3-}}{(175 + 27) \text{ mL}} \times \dfrac{1 \text{ mL}}{10^{-3} \text{ L}} = 0.086633663 \text{ (calc)}$

                        $= 0.0866 \text{ M PO}_4{}^{3-} \text{ (corr)}$

Multi-Concept Problems

15.103 Oxidation numbers of N: N_2O (+1), NO (+2), N_2O_3 (+3), NO_2 (+4), N_2O_5 (+5)

15.105 a) S^{2-} has the maximum number of electrons; it can only lose electrons, not gain them, acting as a reducing agent.

   b) $\text{SO}_4{}^{2-}$ ion, S has a +6 charge. S cannot lose any more electrons; it can only gain them, thereby acting as oxidizing agent.

   c) SO_2, S has charge of +4. S can either lose or gain electrons, acting as reducing agent or oxidizing agent.

   d) SO_3, S has a +6 charge. S cannot lose any more electrons; it can only gain them, thereby acting as oxidizing agent.

15.107 a) +2, +1 b) +3 in both c) +1 in both d) +3 in both

15.109 a) two reduction half-reactions b) one reduction and one oxidation half-reaction
   c) two oxidation half-reactions d) one reduction and one oxidation half-reaction

15.111 1) $5(2 \text{ H}_2\text{O} + \text{PH}_3 \rightarrow \text{H}_3\text{PO}_2 + 4 \text{ H}^+ + 4 \text{ e}^-) + 4(\text{MnO}_4{}^- + 8 \text{ H}^+ + 5 \text{ e}^- \rightarrow \text{Mn}^{2+} + 4 \text{ H}_2\text{O})$
    $10 \text{ H}_2\text{O} + 5 \text{ PH}_3 + 4 \text{ MnO}_4{}^- + 32 \text{ H}^+ + 20 \text{ e}^- \rightarrow 5 \text{ H}_3\text{PO}_2 + 20 \text{ H}^+ + 20 \text{ e}^- + 4 \text{ Mn}^{2+} + 16 \text{ H}_2\text{O}$
      $5 \text{ PH}_3 + 4 \text{ MnO}_4{}^- + 12 \text{ H}^+ \rightarrow 5 \text{ H}_3\text{PO}_2 + 4 \text{ Mn}^{2+} + 6 \text{ H}_2\text{O}$

   2) $2 \text{ H}_2\text{O} + \text{PH}_3 \rightarrow \text{H}_3\text{PO}_2 + 4 \text{ H}^+ + 4 \text{ e}^- + 2(\text{SO}_4{}^{2-} + 4 \text{ H}^+ + 2 \text{ e}^- \rightarrow \text{SO}_2 + 2 \text{ H}_2\text{O})$
    $2 \text{ H}_2\text{O} + \text{PH}_3 + 2 \text{ SO}_4{}^{2-} + 8 \text{ H}^+ + 4 \text{ e}^- \rightarrow \text{H}_3\text{PO}_2 + 4 \text{ H}^+ + 4 \text{ e}^- + 2 \text{ SO}_2 + 4 \text{ H}_2\text{O}$
      $\text{PH}_3 + 2 \text{ SO}_4{}^{2-} + 4 \text{ H}^+ \rightarrow \text{H}_3\text{PO}_2 + 2 \text{ SO}_2 + 2 \text{ H}_2\text{O}$

3) $5(3\,H_2O + As \rightarrow H_3AsO_3 + 3\,H^+ + 3\,e^-) + 3(MnO_4^- + 8\,H^+ + 5\,e^- \rightarrow Mn^{2+} + 4\,H_2O)$
$15\,H_2O + 5\,As + 3\,MnO_4^- + 24\,H^+ + 15\,e^-$
$$\rightarrow 5\,H_3AsO_3 + 15\,H^+ + 15\,e^- + 3\,Mn^{2+} + 12\,H_2O$$
$$5\,As + 3\,MnO_4^- + 3\,H_2O + 9\,H^+ \rightarrow 5\,H_3AsO_3 + 3\,Mn^{2+}$$

4) $2(3\,H_2O + As \rightarrow H_3AsO_3 + 3\,H^+ + 3\,e^-) + 3(SO_4^{2-} + 4\,H^+ + 2\,e^- \rightarrow SO_2 + 2\,H_2O)$
$6\,H_2O + 2\,As + 3\,SO_4^{2-} + 12\,H^+ + 6\,e^- \rightarrow 2\,H_3AsO_3 + 6\,H^+ + 6\,e^- + 3\,SO_2 + 6\,H_2O$
$$2\,As + 3\,SO_4^{2-} + 6\,H^+ \rightarrow 2\,H_3AsO_3 + 3\,SO_2$$

15.113 $4\,Zn + 10\,H^+ + NO_3^-$ $\rightarrow 4\,Zn^{2+} + NH_4^+ + 3\,H_2O$
oxidation:
   Zn $\rightarrow Zn^{2+} + 2\,e^-$
reduction:
   $NO_3^- + 8\,e^-$ $\rightarrow NH_4^+$
   $NO_3^- + 8\,e^- + 10\,H^+$ $\rightarrow NH_4^+$
   $NO_3^- + 10\,H^+ + 8\,e^-$ $\rightarrow NH_4^+ + 3\,H_2O$

15.115 $MnO_4^- + C_2O_4^{2-}$ $\rightarrow Mn^{2+} + CO_2$
oxidation:
   $C_2O_4^{2-}$ $\rightarrow CO_2$
   $C_2O_4^{2-}$ $\rightarrow 2\,CO_2$
   $C_2O_4^{2-}$ $\rightarrow 2\,CO_2 + 2\,e^-$
reduction:
   MnO_4^- $\rightarrow Mn^{2+}$
   $MnO_4^- + 5\,e^-$ $\rightarrow Mn^{2+}$
   $MnO_4^- + 5\,e^- + 8\,H^+$ $\rightarrow Mn^{2+}$
   $MnO_4^- + 5\,e^- + 8\,H^+$ $\rightarrow Mn^{2+} + 4\,H_2O$
combining:
$5(C_2O_4^{2-} \rightarrow 2\,CO_2 + 2\,e^-) + 2(MnO_4^- + 5\,e^- + 8\,H^+ \rightarrow Mn^{2+} + 4\,H_2O)$
$5\,C_2O_4^{2-} + 2\,MnO_4^- + 10\,e^- + 16\,H^+ \rightarrow 10\,CO_2 + 10\,e^- + 2\,Mn^{2+} + 8\,H_2O$
$5\,C_2O_4^{2-} + 2\,MnO_4^- + 16\,H^+ \rightarrow 10\,CO_2 + 2\,Mn^{2+} + 8\,H_2O$ net ionic
$5\,K_2C_2O_4 + 2\,KMnO_4 + 16\,HCl \rightarrow 10\,CO_2 + 2\,MnCl_2 + 8\,H_2O + 12\,KCl$ molecular redox

Answers to Multiple-Choice Practice Test

MC 15.1   c   **MC 15.2**   b   **MC 15.3**   e   **MC 15.4**   c   **MC 15.5**   d   **MC 15.6**   e

MC 15.7   d   **MC 15.8**   d   **MC 15.9**   a   **MC 15.10**   b   **MC 15.11**   d   **MC 15.12**   d

MC 15.13   c   **MC 15.14**   c   **MC 15.15**   d   **MC 15.16**   d   **MC 15.17**   b   **MC 15.18**   a

MC 15.19   e   **MC 15.20**   d

CHAPTER SIXTEEN
Reaction Rates and Chemical Equilibrium

COLLISION THEORY (SEC. 16.1)

16.1 The solute molecules have more motion in the solution, allowing more frequent collisions with other reactant molecules throughout the solution. Only those molecules on the surface of a solid can collide with other reactant molecules.

16.3 X, the slower reaction will have the higher activation energy.

16.5 (1) The combined kinetic energies of the colliding particles must equal or exceed a minimum value, the activation energy, and (2) the orientation of the particles must be favorable.

16.7

 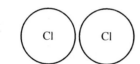

ENDOTHERMIC AND EXOTHERMIC CHEMICAL REACTIONS (SEC. 16.2)

16.9
a) $2 SO_2 + O_2 \rightarrow 2 SO_3 + heat$
   Exothermic reactions give off heat.
b) $N_2 + O_2 + heat \rightarrow 2 NO$
   Endothermic reactions absorb heat.
c) $CH_4 + 2 O_2 \rightarrow CO_2 + 2 H_2O + heat$
   Exothermic reactions give off heat.
d) $CaCO_3 + heat \rightarrow CaO + CO_2$
   Endothermic reactions absorb heat.

16.11 a) endothermic   b) exothermic   c) exothermic   d) endothermic

16.13
a) (1) The average energy of the reactant molecules is less than the average energy of the products.
b) (3) The average energy of the reactant molecules is greater than the average energy of the products.
c) (3) The average energy of the reactant molecules is greater than the average energy of the products.
d) (1) The average energy of the reactant molecules is less than the average energy of the products.

16.15

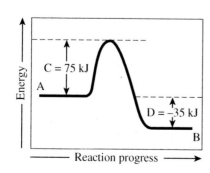

a) average energy of the reactants = A
b) average energy of the products = B
c) activation energy = C
d) energy liberated = D

16.17

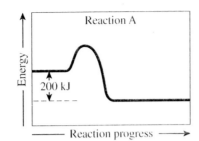

Similarities: Both reactions are exothermic and the energy difference between reactants and products is the same in both reactions. Differences: The activation energy is lower in the reaction that takes place at room temperature, Reaction A.

FACTORS THAT INFLUENCE CHEMICAL REACTION RATES (SEC. 16.3)

16.19   a) A temperature change causes the collision frequency to change and also causes the molecular energy to change.
       b) A catalyst provides an alternate pathway with a lower activation energy.

16.21   For the reaction: $Fe(s) + 2HCl(aq) \rightarrow FeCl_2(aq) + H_2(g)$
       a) increasing the temperature   (1) increases the rate of reaction
       b) solution is stirred   (1) increases the rate of reaction
       c) increasing the concentration of acid, HCl   (1) increases the rate of reaction
       d) Fe is ground into powder   (1) increases the rate of reaction

16.23   a) 1—The lower the activation energy, the faster the reaction.
       b) 3—The higher the temperature, the faster the reaction.
       c) 4—The higher the concentration of a reactant, the faster the reaction.
       d) 3—The higher the temperature and the lower the activation energy, the faster the reaction.

16.25   Diagrams are the same except for the magnitude of the activation energy. A catalyst lowers the activation energy.

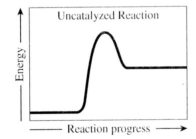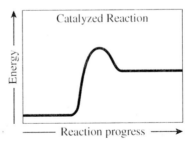

CHEMICAL EQUILIBRIUM (SEC. 16.4)

16.27   The rate of the forward reaction must equal the rate of the reverse reaction.

16.29   a) Forward reaction: $CO(g) + H_2O(g) \rightarrow CO_2(g) + H_2(g)$
       b) Reverse reaction: $CO_2(g) + H_2(g) \rightarrow CO(g) + H_2O(g)$

16.31 $X \rightleftharpoons Y$
       Diagram I: 10 X   Diagram II: 7 X, 3 Y   Diagram III: 6 X, 4 Y   Diagram IV: 6 X, 4 Y
       The system changes in Diagram II; since Diagrams III and IV are the same, the system has possibly reached equilibrium.

16.33   Consider the balanced reaction: A_2 + 2 B → 2 AB
Diagram I: 6 B, 4 A_2 Diagram II: 2 B, 2 A_2, 4 AB
Diagram III: 4 B, 2 A_2, 4 AB   Diagram IV: 4 B, 3 A_2, 2 AB
Diagrams II and IV represent a possible equilibrium state; they remain balanced.

16.35   As the amount of reactants decreases, the amount of products formed increases.

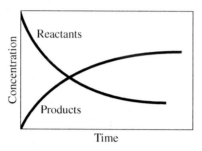

EQUILIBRIUM MIXTURE STOICHIOMETRY (SEC. 16.5)

16.37

	2 SO$_3$	⇌	2 SO$_2$	+	O$_2$
start	0.0200 mole		0 mole		0 mole
change	?		?		?
equilibrium	?		?		0.0029 mole

The change in O_2 = +0.0029 moles to get from 0 at start to 0.0029 at equilibrium.
The change in SO_2 must be twice as great as the change in O_2 or +0.0058 mole (produced).
The change in SO_3 must be twice as great as the change in O_2 or −0.0058 mole (consumed).
At equilibrium, O_2 is 0.0029 moles (given).
At equilibrium, SO_2 is 0 + 0.0058 mole (start + change).
At equilibrium, SO_3 is 0.0200 − 0.0058 = 0.0142 mole (calc and corr).
The full table becomes:

	2 SO$_3$	⇌	2 SO$_2$	+	O$_2$
start	0.0200 mole		0 mole		0 mole
change	−0.0058 mole		+0.0058 mole		0.0029 mole
equilibrium	0.0142 mole		0.0058 mole		0.0029 mole

16.39

	2 NH$_3$(g)	⇌	N$_2$(g)	+	3 H$_2$(g)
start	0.296 mole		0.170 mole		0.095 mole
change	−0.028 mole		+0.014 mole		+0.042 mole
equilibrium	0.268 mole		0.184 mole		0.137 mole

16.41

$$2\,CO(g) \quad + \quad O_2(g) \quad \rightleftharpoons \quad 2\,CO_2(g)$$

start	0.100 mole	0.200 mole	0 mole
change	−0.006 mole	−0.003 mole	+0.006 mole
equilibrium	0.094 mole	0.197 mole	0.006 mole

EQUILIBRIUM CONSTANTS (SEC. 16.6)

16.43   a) $H_2(g) + CO_2(g) \rightleftharpoons H_2O(g) + CO(g)$ $K_{eq} = \dfrac{[H_2O][CO]}{[H_2][CO_2]}$

b) $H_2(g) + Cl_2(g) \rightleftharpoons 2\,HCl(g)$ $K_{eq} = \dfrac{[HCl]^2}{[H_2][Cl_2]}$

c) $4\,NO(g) + 6\,H_2O(g) \rightleftharpoons 4\,NH_3(g) + 5\,O_2(g)$ $K_{eq} = \dfrac{[NH_3]^4[O_2]^5}{[NO^4][H_2O]^6}$

d) $CH_4(g) + 2\,O_2(g) \rightleftharpoons CO_2(g) + 2\,H_2O(g)$ $K_{eq} = \dfrac{[CO]_2[H_2O]^2}{[CH_4][O_2]^2}$

16.45   a) $4\,Al(s) + 3\,O_2(g) \rightleftharpoons 2\,Al_2O_3(s)$ $K_{eq} = \dfrac{1}{[O_2]^3}$

b) $CaCO_3(s) \rightleftharpoons CaO(s) + CO_2(g)$ $K_{eq} = [CO_2]$

c) $NaCl(aq) + AgNO_3(aq) \rightleftharpoons AgCl(s) + NaNO_3(aq)$ $K_{eq} = \dfrac{[NaNO_3]}{[NaCl][AgNO_3]}$

d) $3\,CuO(s) + 2\,NH_3(g) \rightleftharpoons 3\,Cu(s) + N_2(g) + 3\,H_2O(g)$ $K_{eq} = \dfrac{[N_2][H_2O]^3}{[NH_3]^2}$

16.47   a) $K_{eq} = \dfrac{[B]^2[C]}{[A]} = \dfrac{(2.00)^2(5.00)}{3.00} = 6.6666667\ (calc) = 6.67\ (corr)$

b) $K_{eq} = \dfrac{[C]^2}{[A][B]^3} = \dfrac{(5.00)^2}{(3.00)(2.00)^3} = 1.0416667\ (calc) = 1.04\ (corr)$

c) $K_{eq} = \dfrac{[A][C]}{[B]^2} = \dfrac{(3.00)(5.00)}{(2.00)^2} = 3.75\ (calc\ and\ corr)$

d) $K_{eq} = \dfrac{[A]^3}{[C]^4[B]} = \dfrac{(3.00)^3}{(5.00)^4(2.00)} = 0.0216\ (calc\ and\ corr)$

16.49   Diagram IV has the largest product to reactant ratio: $K_{eq} = \dfrac{[AB]^2}{[A_2][B_2]} = 64$

16.51   Diagram I has the only product to reactant ratio of 64: $K_{eq} = \dfrac{[AB]^2}{[A_2][B_2]} = 64$

$64\,([A_2][B_2]) = [AB]^2$ If $AB = 8$ and $A_2 = 1$ and $B_2 = 1$

16.53 $K_{eq} = \dfrac{[CH_4][H_2S]^2}{[CS_2][H_2]^4}$; or $0.0280 = \dfrac{(0.00100)(1.43)^2}{[CS_2](1.00)^4}$

Solving for $[CS_2]$ gives: $[CS_2] = \dfrac{(0.00100)(1.43)^2}{(0.0280)(1.00)^4} = 0.073032143$ (calc) $= 0.0730$ M CS_2 (corr)

16.55 Find the molar concentration of each gas by dividing the moles by the volume,

$$K_{eq} = \frac{[PCl_5]}{[PCl_3][Cl_2]} = \frac{\left(\dfrac{0.0189}{6.00}\right)}{\left(\dfrac{0.0222}{6.00}\right)\left(\dfrac{0.1044}{6.00}\right)} = 48.92823858 \text{ (calc)} = 48.9 \text{ (corr)}$$

16.57 For the reaction: $2 A(g) \rightleftharpoons 2 B(g) + C(g)$

At temperature T, $K_{eq} = \dfrac{[B]^2[C]}{[A]^2} = 2.4 \times 10^{-3}$

If the [B] is found to be 2 times that of [A], what is the concentration of C?
Let the concentration of A = X, and the concentration of B = 2X.

$2.4 \times 10^{-3} = \dfrac{[2X]^2[C]}{[X]^2}$ $[C] = \dfrac{X^2 \times (2.4 \times 10^{-3})}{4X^2} = 6 \times 10^{-4}$(calc) $= 6.0 \times 10^{-4}$(corr)

16.59   a) mostly reactants   b) mostly products
c) mostly products   d) significant amounts of both reactants and products

LE CHÂTELIER'S PRINCIPLE (SEC. 16.9)

16.61 For the reaction: $A(g) + B(g) \rightleftharpoons C(g) + D(g)$
a) increasing the concentration of A drives the equilibrium toward the products, and the concentration of D will subsequently increase (1).
b) decreasing the concentration of B drives the equilibrium toward the reactants, and the concentration of D will subsequently decrease (2).
c) increasing the concentration of C drives the equilibrium toward the reactants, and the concentration of D will subsequently decrease (2).
d) decreasing the concentration of C drives the equilibrium toward the products, and the concentration of D will subsequently increase (1).

16.63   a) Increasing the concentration of the reactants shifts the equilibrium toward the products.
b) Decreasing the concentration of the reactant will shift the equilibrium toward the reactants.
c) Since this reaction is exothermic, we treat heat as a product. Increasing the temperature will shift the reaction toward the reactants.
d) Increasing the pressure will shift the equilibrium toward the least number of moles of gas, in this case toward the products with 6 moles of gas versus the reactants with 7 moles of gas.

16.65   a) Since this reaction is endothermic, we treat heat as a reactant. Heating the reaction mixture will shift the reaction toward the products.
b) Increasing the concentration of the products shifts the equilibrium toward the reactants.
c) Increasing the pressure by adding an inert gas will have no effect on the equilibrium.
d) Increasing the size of the container will shift the reaction toward the side with the most gas molecules. In this case it shifts toward the products with 6 moles of gas versus the reactants with 4 moles of gas.

16.67 Increased product formation is a shift to the right. Higher temperatures will shift endothermic reactions to the right, so only in c) would high temperature favor product formation.

16.69 Increasing the temperature produces more products; therefore the reaction is endothermic.

Multi-Concept Problems

16.71   a) $K_{eq} = \dfrac{[0.350]^2[0.190]}{[0.0300]^2} = 25.861111$ (calc) $= 25.9$ (corr); system is at equilibrium.

     b) $K_{eq} = \dfrac{[0.700]^2[0.380]}{[0.0600]^2} = 51.722222$ (calc) $= 51.7$ (corr)

        K_{eq} value is too high for equilibrium; a shift to the left will reduce the size of K_{eq}.

     c) $K_{eq} = \dfrac{[0.356]^2[0.160]}{[0.0280]^2} = 25.864489$ (calc) $= 25.9$ (corr); system is at equilibrium.

     d) $K_{eq} = \dfrac{[0.330]^2[0.180]}{[0.0100]^2} = 196.02$ (calc) $= 196$ (corr)

        K_{eq} value is too high for equilibrium; a shift to the left will reduce the size of K_{eq}.

16.73 Since the numerator and denominator are reversed, the second K is the reciprocal of the first K.

$$K_{\text{reverse reaction}} = \frac{1}{K_{\text{forward reaction}}} = \frac{1}{2 \times 10^3} = 0.0005 \text{ (calc and corr)}$$

16.75   a) no change   b) no change
        c) change in equilibrium constant value   d) no change

16.77   a) right   b) no change
        c) left   d) right

16.79 For the reaction: $A(g) + B(g) \rightleftharpoons C(g) + D(g)$

$$K_{eq} = \frac{[C][D]}{[A][B]} \text{ and } [D] = \frac{K_{eq} \times [A][B]}{[C]}$$

     1. initial concentrations of $[A] = 1.2$ and $[B] = 0.8$

$$[D] = \frac{K_{eq} \times 1.2 \times 0.8}{[C]} = \frac{K_{eq} \times 0.96}{[C]}$$

     2. initial concentrations of $[A] = 1.2$ and $[B] = 0.8$ and $[C] = 0.5$

$$[D] = \frac{K_{eq} \times 1.2 \times 0.8}{0.5} = K_{eq} \times 1.92$$

     3. initial concentrations of $[A] = 2.2$ and $[B] = 0.8$

$$[D] = \frac{K_{eq} \times 2.2 \times 0.8}{[C]} = \frac{K_{eq} \times 1.76}{[C]}$$

4. initial concentrations of $[A] = 2.2$ and $[B] = 1.8$

$$[D] = \frac{K_{eq} \times 2.2 \times 1.8}{[C]} = \frac{K_{eq} \times 3.96}{[C]}$$

a) for which of these situations is [D] the least at equilibrium?

Assuming $K_{eq} = 1$, the lower the concentration of A and B at equilibrium, the lower the concentration of D will be at equilibrium. (1) will have the least impact on the concentration of D at equilibrium.

b) for which of these situations is [D] the greatest at equilibrium?

Assuming $K_{eq} = 1$, increasing the concentration of A and B drives the equilibrium toward the products, and therefore increases the concentration of D at equilibrium. (4) will increase the concentration of D the most at equilibrium.

Answers to Multiple-Choice Practice Test

MC 16.1 c	MC 16.2 c	MC 16.3 c	MC 16.4 e	MC 16.5 e
MC 16.6 b	MC 16.7 b	MC 16.8 d	MC 16.9 e	MC 16.10 b
MC 16.11 a	MC 16.12 d	MC 16.13 d	MC 16.14 b	MC 16.15 b
MC 16.16 e	MC 16.17 b	MC 16.18 c	MC 16.19 b	MC 16.20 c